ASHEVILLE-BUNCOMBE TECHNICAL INSTITUTE

NORTH CAROLINA
STATE
DEPT.

DISCARDED

NOV 2 0 2024

HANDBOOK OF THE
ATOMIC ELEMENTS

HANDBOOK
OF THE
ATOMIC ELEMENTS

by

R. A. WILLIAMS

PHILOSOPHICAL LIBRARY

NEW YORK

Copyright, © 1970, by Philosophical Library, Inc.
15 East 40th Street, New York, N.Y. 10016

All Rights Reserved

Library of Congress Catalog Card No. 78-92089

SBN: 8022–2340–0

Manufactured in the United States of America

For Marta

CONTENTS

INTRODUCTION

The purpose of this book is manifold. Among the first reasons for putting together such a book is the belief that once we have measured something and are able to express the result in numbers, we can certainly say that we know something about that which we have just measured. We are also able to perform the same measurement on different members in the same category and compare the result—the number that is obtained from this last measurement with that gotten from the first. The result is, hopefully, that more will be known about both.

Another reason is a pragmatic one. Not only must one measure a quantity and express it as a recorded number; also the results must be easily available for rapid reference in the future.

The *Handbook of the Atomic Elements* is specifically designed to contain the most important, or at least the most often used, measurements or parameters pertinent to the elements and their classification into periodic categories.

The elements, ranging from Hydrogen at atomic number 1, through Lawrencium at atomic number 103, are arranged in alphabetical order so that any element can be located in less time than it takes to look up a word in a dictionary.

The information on each of the elements is based on Carbon-12 and values released by the International Union of Pure and Applied Chemistry. The physical values are grouped for ready use and include the following coverage of data: the name of the element and its atomic symbol; the principal quantum no., n(period); atomic number; atomic weight; x-ray-notation; group; category; state; origin; the number of protons and electrons; the valence (s); the subshell filling; atomic, covalent and ionic radii; the ground state electron configuration; valence electrons; core member; and acid-base properties. Also included are the atomic volume; density; type of crystal structure; the electronegativity and electrical conductance; first ionization energy; ionization potential; heat of vaporization and fusion; and specific heat. Also, thermal conductance, boiling point and melting point. Radioactive isotopes are listed and their half-lives accompanied by the emission particle.

9

All of the values given are provided with units and dimensional analysis.

Numerous helpful tables and illustrations and a copy of the periodic classifications of the elements (chart) are included.

Students, or anyone else having reason to know something about the elements, should find this book to be very helpful as a constant and handy reference volume.

The author of *Handbook of the Atomic Elements* wishes to extend sincere gratitude to those persons who have offered many helpful comments regarding the preliminary editions of this book. But I must not conclude here without due mention that the staff of Philosophical Library have given me assistance in many ways that go far beyond the call of a publisher's duties.

R. A. W.
Los Angeles, December, 1969

NOTES ON THE TEXT

The elements are arranged in alphabetical order for ease of location. The name of each element will be found in bold face at the top of the page, and is followed on the same line by its atomic symbol and, in applicable cases, by the derived name in parentheses.

Atomic weights followed by the symbol △ are either not officially accepted at this time, or are the weight of the isotope with the longest known half-life or, as in the cases of Berkelium, Californium, Lawrencium, Polonium, Technetium and Promethium, are the weight of the most common isotope.

The first category for each element is the Principal Quantum Number. In all cases the quality "n. (period)" is assumed.

Where there is no data available for a particular category, a stroke is used.

Under the category "State," the physical state for the element is recorded for that element at 30° C and 1 atmosphere.

11

Actinium   Ac

Principal Quantum No.	6	Valence Electrons	$6d^1$ $7s^2$ (+3)	
Atomic No.	89	Acid-Base Property	/	
Atomic Wt.	227 Δ	Density	/	
X-Ray Notation	P	Crystal Structure	/	
Group	IV B	Electro-negativity	1.1	
Category	Actinon	Electrical Conductance	/	
State	Solid			
Origin	Natural			
No. of Protons	89	First Ionization Energy	/	
No. of Electrons	89	Ionization Potential	/	
Valence	+3	Heat of Vaporization	/	
Subshell Filling	d	Heat of Fusion	/	
Atomic Radius	/	Specific Heat	/	
Covalent Radius	/	Thermal Conductance	/	
Ionic Radius	1.18 Å (+3)	Boiling Point	/	
Atomic Vol.	/	Melting Point	1050° C	
Core	Krypton			

Ground State Electron Config.
$1s^2$ $2s^2$ $2p^6$ $3s^2$ $3p^6$ $3d^{10}$
$4s^2$ $4p^6$ $4d^{10}$ $4f^{14}$ $5s^2$ $5p^6$
$5d^{10}$ $5f^0$ $5g^0$ $6s^2$ $6p^6$ $6d^1$
$7s^2$

Radioactive Isotopes
Ac (at.wt.=227); occurs naturally;
½life=22 years with decay via α and
$β^-$ (particles).

Aluminum   Al

Principal Quantum No.	3	Valence Electrons	$3s^2\ 3p^1\ (+3)$
Atomic No.	13	Acid-Base Property	Amphoteric
Atomic Wt.	26.9815	Density	2.70 gm/ml
X-Ray Notation	M	Crystal Structure	Cubic—face centered
Group	III A	Electronegativity	1.5
Category	Metal	Electrical Conductance	0.382 micro-ohms
State	Solid		
Origin	Natural		
No. of Protons	13	First Ionization Energy	138 Kcal/gm-mole
No. of Electrons	13	Ionization Potential	6.0 ev
Valence	+3	Heat of Vaporization	67.9 Kg-cal/gm atom
Subshell Filling	p	Heat of Fusion	2.55 Kg-cal/gm atom
Atomic Radius	1.43 Å	Specific Heat	0.215 cal/gm/°C
Covalent Radius	1.18 Å	Thermal Conductance	0.50 cal/cm^2/cm/°C/sec
Ionic Radius	0.50 Å	Boiling Point	2450° C
Atomic Vol.	10.0 W/D	Melting Point	660° C
Core	Neon		

Ground State Electron Config.
$1s^2\ 2s^2\ 2p^6\ 3s^2\ 3p^1$

Radioactive Isotopes
None

Americium   Am

Principal Quantum No.	7	Valence Electrons	$5f^7$ $6d^0$ $7s^2$
Atomic No.	95	Acid-Base Property	/
Atomic Wt.	243 Δ		
X-Ray Notation	Q	Density	11.7 gm/ml
Group	VIII B	Crystal Structure	/
Category	Actinon	Electro-negativity	/
State	Solid	Electrical Conductance	0.007 micro-ohms
Origin	Synthetic		
No. of Protons	95	First Ionization Energy	/
No. of Electrons	95	Ionization Potential	/
Valence	+2, +3, +4, +5, +6	Heat of Vaporization	/
Subshell Filling	f	Heat of Fusion	/
Atomic Radius	/	Specific Heat	0.033 cal/gm/°C
Covalent Radius	/	Thermal Conductance	/
Ionic Radius	1.06 Å(+3); 0.85 Å(+4)	Boiling Point	/
		Melting Point	/
Atomic Vol.	20.8 W/D		
Core	Krypton		

Ground State Electron Config.
$1s^2$ $2s^2$ $2p^6$ $3s^2$ $3p^6$ $3d^{10}$
$4s^2$ $4p^6$ $4d^{10}$ $4f^{14}$ $5s^2$ $5p^6$
$5d^{10}$ $5f^7$ $5g^0$ $6s^2$ $6p^6$ $6d^0$
$7s^2$

Radioactive Isotopes
Am (at.wt.=241); ½life=470 years; with decay via α, γ and e⁻.
Am (at.wt.=242); ½life=100 years with decay via α, β, γ and K.
Am (at.wt.=243); ½life=8000 years with decay via α and γ.
Radioactivity is induced in all isotopes.

Antimony — Sb (Stibium)

Principal Quantum No.	5		Valence Electrons	$4d^{10}\ 5s^2\ 5p^3$
Atomic No.	51		Acid-Base Property	Moderately acidic
Atomic Wt.	121.75		Density	6.62 gm/ml
X-Ray Notation	0		Crystal Structure	Rhombohedral
Group	V A		Electronegativity	1.9
Category	Transitional metal		Electrical Conductance	0.026 micro-ohms
State	Solid		First Ionization Energy	199 Kcal/gm-mole
Origin	Natural		Ionization Potential	8.6 ev
No. of Protons	51		Heat of Vaporization	46.6 Kg-cal/gm-atom
No. of Electrons	51		Heat of Fusion	4.74 Kg-cal/gm-atom
Valence	+3, +5 (also, −3, +4)		Specific Heat	0.049 cal/gm/°C
Subshell Filling	p		Thermal Conductance	0.05 cal/cm²/cm/°C/sec
Atomic Radius	1.59 Å		Boiling Point	1380° C
Covalent Radius	1.38 Å		Melting Point	630° C
Ionic Radius	2.45 Å(−3); 0.62 Å(+5)			
Atomic Vol.	18.4 W/D			
Core	Krypton			

Ground State Electron Config.
$1s^2\ 2s^2\ 2p^6\ 3s^2\ 3p^6\ 3d^{10}$
$4s^2\ 4p^6\ 4d^{10}\ 4f^0\ 5s^2\ 5p^3$

Radioactive Isotopes
Sb (at.wt.=122); ½life=2.8 days with decay via β^-, β^+, γ and K.
Sb (at.wt.=124); ½life=60 days with decay via β^- and γ.
Radioactivity is induced.

Argon — Ar

Principal Quantum No.	3	Valence Electrons	$3s^2\ 3p^6$ (0)
Atomic No.	18	Acid-Base Property	/
Atomic Wt.	39.948	Density	1.40 gm/ml
X-Ray Notation	M	Crystal Structure	Cubic—face centered
Group	Inert Gases	Electronegativity	/
Category	Inert Gas		
State	Gas	Electrical Conductance	/
Origin	Natural		
No. of Protons	18	First Ionization Energy	363
No. of Electrons	18	Ionization Potential	15.8 ev
Valence	0	Heat of Vaporization	1.56 Kg-cal/gm-atom
Subshell Filling	p	Heat of Fusion	0.281 Kg-cal/gm-atom
Atomic Radius	/	Specific Heat	0.125 cal/gm/°C
Covalent Radius	1.74 Å	Thermal Conductance	0.00004 cal/cm²/cm/°C/sec
Ionic Radius	/	Boiling Point	−185.8° C
Atomic Vol.	24.2 W/D	Melting Point	−189.5° C
Core	/		

Ground State Electron Config.
$1s^2\ 2s^2\ 2p^6\ 3s^2\ 3p^6$

Radioactive Isotopes
None

Arsenic
As

Principal Quantum No.	4	Valence Electrons	$3d^{10}$ $4s^2$ $4p^3$
Atomic No.	33	Acid-Base Property	Moderately acidic
Atomic Wt.	74. 9216	Density	5.72 gm/ml
X-Ray Notation	N	Crystal Structure	Rhombohedral
Group	V A	Electro-negativity	2.0
Category	Non-metal (or semi-metal)	Electrical Conductance	0.029 micro-ohm
State	Solid	First Ioniza-tion Energy	231 Kcal/gm-mole
Origin	Natural		
No. of Protons	33	Ionization Potential	10 ev
No. of Electrons	33	Heat of Vaporization	7.75 Kg-cal/gm-atom
Valence	+3, +5 (also, −3, +2)	Heat of Fusion	6.62 Kg-cal/gm-atom
Subshell Filling	p	Specific Heat	0.082 cal/gm/°C
Atomic Radius	1.39 Å	Thermal Conductance	/
Covalent Radius	1.19 Å	Boiling Point	613° C
Ionic Radius	2.22 Å(−3); 0.47 Å(+5)	Melting Point	817° C
Atomic Vol.	13.1 W/D		
Core	Argon		

Ground State Electron Config.
$1s^2$ $2s^2$ $2p^6$ $3s^2$ $3p^6$ $3d^{10}$ $4s^2$ $4p^3$

Radioactive Isotopes
As (at.wt.=76); ½life=26.7 hours with decay via β^- and γ.
As (at.wt.=77); ½life=39 hours with decay via β^- and γ.
Radioactivity is induced.

Astatine   At

Principal Quantum No.	6	Valence Electrons	$4f^{14}\ 5d^{10}\ 6s^2\ 6p^5$
Atomic No.	85	Acid-Base Property	/
Atomic Wt.	210	Density	/
X-Ray Notation	P	Crystal Structure	/
Group	VII A	Electronegativity	2.2
Category	Halogen		
State	Solid	Electrical Conductance	/
Origin	Natural		
No. of Protons	85	First Ionization Energy	/
No. of Electrons	85	Ionization Potential	/
Valence	−1	Heat of Vaporization	8 Kg-cal/gm-atom
Subshell Filling	p	Heat of Fusion	/
Atomic Radius	/	Specific Heat	/
Covalent Radius	/	Thermal Conductance	/
Ionic Radius	/	Boiling Point	/
Atomic Vol.	/	Melting Point	302° C
Core	Krypton		

Ground State Electron Config.
$1s^2\ 2s^2\ 2p^6\ 3s^2\ 3p^6\ 3d^{10}$
$4s^2\ 4p^6\ 4d^{10}\ 4f^{14}\ 5s^2\ 5p^6$
$5d^{10}\ 5f^0\ 5g^0\ 6s^2\ 6p^5$

Radioactive Isotopes
None

19

Barium   Ba

Principal Quantum No.	6		Valence Electrons	$6s^2$
Atomic No.	56		Acid-Base Property	Strongly basic
Atomic Wt.	137.34		Density	3.5 gm/ml
X-Ray Notation	P		Crystal Structure	Cubic—body centered
Group	II A		Electro-negativity	0.9
Category	Alkali earth metal		Electrical Conductance	0.016 micro-ohm
State	Solid		First Ioniza-tion Energy	120 kcal/gm-mole
Origin	Natural		Ionization Potential	5.2 ev
No. of Protons	56		Heat of Vaporization	35.7 Kg-cal/gm-atom
No. of Electrons	56		Heat of Fusion	1.83 Kg-cal/gm-atom
Valence	+2		Specific Heat	0.068 cal/gm/°C
Subshell Filling	s		Thermal Conductance	/
Atomic Radius	2.22 Å		Boiling Point	1640° C
Covalent Radius	1.98 Å		Melting Point	714° C
Ionic Radius	1.35 Å(+2)			
Atomic Vol.	39 W/D			
Core	Krypton			

Ground State Electron Config.
$1s^2$ $2s^2$ $2p^6$ $3s^2$ $3p^6$ $3d^{10}$
$4s^2$ $4p^6$ $4d^{10}$ $4f^0$ $5s^2$ $5p^6$
$5d^0$ $5f^0$ $5g^0$ $6s^2$

Radioactive Isotopes
Ba (at.wt.=131); ½life=12 days
    with decay via γ and K.
Ba (at.wt.=133); ½life=7.5 days
    with decay via γ, e^- and K.
Radioactivity is induced.

Berkelium — Bk

Principal Quantum No.	7	Valence Electrons	$5f^7$ $6d^2$ $7s^2$
Atomic No.	97	Acid-Base Property	/
Atomic Wt.	247 Δ		
X-Ray Notation	Q	Density	/
Group	II B	Crystal Structure	/
Category	Actinon	Electro-negativity	/
State	Solid		
Origin	Synthetic	Electrical Conductance	/
No. of Protons	97	First Ionization Energy	/
No. of Electrons	97	Ionization Potential	/
Valence	+3, +4	Heat of Vaporization	/
Subshell Filling	f	Heat of Fusion	/
Atomic Radius	/	Specific Heat	/
Covalent Radius	/	Thermal Conductance	/
Ionic Radius	/	Boiling Point	/
Atomic Vol.	/	Melting Point	/
Core	Krypton		

Ground State Electron Config.
$1s^2$ $2s^2$ $2p^6$ $3s^2$ $3p^6$ $3d^{10}$ $4s^2$
$4p^6$ $4d^{10}$ $4f^{14}$ $5s^2$ $5p^6$ $5d^{10}$ $5f^8$
$5g^0$ $6s^2$ $6p^6$ $6d^1$ $7s^2$

Radioactive Isotopes
Bk (at.wt. 245); ½life=4.9 days
with decay via α, γ and K.
Bk (at.wt. 249); ½life=291 days
with decay via α and β; spontaneous
fission.
Radioactivity is induced.

Beryllium   Be

Principal Quantum No.	2		Valence Electrons	2S² (+2)
Atomic No.	4		Acid-Base Property	Amphoteric
Atomic Wt.	9.0122		Density	1.86 gm/ml
X-Ray Notation	L		Crystal Structure	Hexagonal
Group	II A		Electro-negativity	1.5
Category	Alkali earth metal		Electrical Conductance	0.25 micro-ohm
State	Solid		First Ioniza-tion Energy	215 kcal/gm-mole
Origin	Natural			
No. of Protons	4		Ionization Potential	9.3 ev
No. of Electrons	4		Heat of Vaporization	73.9 Kg-cal/gm-atom
Valence	+2			
Subshell Filling	s		Heat of Fusion	2.8 Kg-cal/gm-atom
Atomic Radius	1.12 Å		Specific Heat	0.45 cal/gm/°C
Covalent Radius	0.90 Å		Thermal Conductance	0.38 cal/cm²/cm/°C/sec
Ionic Radius	0.31 Å		Boiling Point	2770° C
Atomic Vol.	5.0 W/D		Melting Point	1277° C
Core	/			

Ground State Electron Config.   Radioactive Isotopes
1s² 2s²   None

Bismuth   Bi

Principal Quantum No.	6	Valence Electrons	6P³ (+3)
Atomic No.	83	Acid-Base Property	Moderately acidic
Atomic Wt.	208.980	Density	9.8 gm/ml
X-Ray Notation	P	Crystal Structure	Rhombohedral
Group	V A	Electronegativity	1.9
Category	Metal	Electrical Conductance	0.009 micro-ohm
State	Solid		
Origin	Natural	First Ionization Energy	185 kcal/gm-mole
No. of Protons	83	Ionization Potential	8 ev
No. of Electrons	83	Heat of Vaporization	42.7 Kg-cal/gm-atom
Valence	+3 (also, −3, +2, +4, +5)	Heat of Fusion	2.6 Kg-cal/gm-atom
Subshell Filling	p	Specific Heat	0.034 cal/gm/°C
Atomic Radius	1.70 Å	Thermal Conductance	0.02 cal/cm²/cm/°C/sec
Covalent Radius	1.46 Å	Boiling Point	1560° C
Ionic Radius	1.20 Å(+3); 0.74 Å(+5)	Melting Point	271° C
Atomic Vol.	21.3 W/D		
Core	Krypton		

Ground State Electron Config.
1s² 2s² 2p⁶ 3s² 3p⁶ 3d¹⁰
4s² 4p⁶ 4d¹⁰ 4f¹⁴ 5s² 5p⁶
5d¹⁰ 5f⁰ 5g⁰ 6s² 6p³

Radioactive Isotopes
Bi (at.wt.=210); ½life=5 days
with decay via α and β⁻.
Radioactive isotopes occur naturally.

Boron   B

Principal Quantum No.	2	Valence Electrons	$2s^2\ 2p^1\ (+3)$
Atomic No.	5	Acid-Base Property	Moderately acidic
Atomic Wt.	10.81	Density	2.34 gm/ml
X-Ray Notation	L	Crystal Structure	Hexagonal
Group	III A		
Category	Semi-metal	Electro-negativity	2.0
State	Solid	Electrical Conductance	10^{-12} micro-ohm
Origin	Natural		
No. of Protons	5	First Ioniza-tion Energy	191 kcal/gm-mole
No. of Electrons	5	Ionization Potential	8.3 ev
Valence	+3, −3	Heat of Vaporization	75 Kg-cal/gm-atom
Subshell Filling	p	Heat of Fusion	5.3 Kg-cal/gm-atom
Atomic Radius	0.98 Å	Specific Heat	0.309 cal/gm/°C
Covalent Radius	0.82 Å	Thermal Conductance	/
Ionic Radius	0.02 Å(+3)	Boiling Point	/
Atomic Vol.	4.6 W/D	Melting Point	/
Core	/		

Ground State Electron Config.
$1s^2\ 2s^2\ 2p^1$

Radioactive Isotopes
None

Bromine   Br

Principal Quantum No.	4	Valence Electrons	$3d^{10}$ $4s^2$ $4p^5$
Atomic No.	35	Acid-Base Property	Very Acidic
Atomic Wt.	79.90363	Density	3.12 gm/ml
X-Ray Notation	N	Crystal Structure	Cubic
Group	VII A	Electro-negativity	2.8
Category	Halogen	Electrical Conductance	10^{-18} micro-ohm
State	Liquid (solid crystals)	First Ioniza-tion Energy	273 kcal/gm-mole
Origin	Natural	Ionization Potential	11.8 ev
No. of Protons	35	Heat of Vaporization	3.58 kg-cal/gm-atom
No. of Electrons	35	Heat of Fusion	1.26 kg-cal/gm-atom
Valence	−1, +1 (also, +3, +4, +5)	Specific Heat	0.070 cal/gm/°C
Subshell Filling	p	Thermal Conductance	/
Atomic Radius	/	Boiling Point	58° C
Covalent Radius	1.14 Å	Melting Point	−7.2° C
Ionic Radius	1.95 Å(−1); 0.39 Å(+5)		
Atomic Vol.	23.5 W/D		
Core	Argon		

Ground State Electron Config.
1s² 2s² 2p⁶ 3s² 3p⁶ 3d¹⁰
4s² 4p⁵

Radioactive Isotopes
Br (at.wt.=82); ½life=36 hours
with decay via β^- and γ.

Cadmium   Cd

Principal Quantum No.	5	Valence Electrons	$5s^2$
Atomic No.	48	Acid-Base Property	Moderately basic
Atomic Wt.	112.40	Density	8.65 gm/ml
X-Ray Notation	Q	Crystal Structure	Hexagonal
Group	II B	Electro-negativity	1.7
Category	Heavy transitional metal	Electrical Conductance	0.146 micro-ohm
State	Solid	First Ioniza-tion Energy	207 kcal/gm-mole
Origin	Natural		
No. of Protons	48	Ionization Potential	9.0 ev
No. of Electrons	48	Heat of Vaporization	23.9 Kg-cal/gm-atom
Valence	+2 (also, +1)		
Subshell Filling	s	Heat of Fusion	1.46 Kg-cal/gm-atom
Atomic Radius	1.54 Å	Specific Heat	0.055 cal/gm/°C
Covalent Radius	1.48 Å	Thermal Conductance	0.22 cal/cm²/cm/°C/sec
Ionic Radius	0.97 Å(+2)	Boiling Point	765° C
Atomic Vol.	13.1 W/D	Melting Point	320.9° C
Core	Krypton		

Ground State Electron Config.
1s² 2s² 2p⁶ 3s² 3p⁶ 3d¹⁰
4s² 4p⁶ 4d¹⁰ 4f⁰ 5s²

Radioactive Isotopes
Cd (at.wt.=115); ½life=43 days
with decay via β^- and γ.
Radioactivity is induced.

Date Due Slip

Tech Community College
9/19 01:07PM

* * * * * * * * * * * * * * * * * * * *

RON: *********8003

33312000106088
DATE: 08/05/19
book of the atomic elements /

NO. QD 466 .W52 1970

Calcium   Ca

Principal Quantum No.	4	Valence Electrons	$4s^2$
Atomic No.	20	Acid-Base Property	Strongly basic
Atomic Wt.	40.08	Density	1.55 gm/ml
X-Ray Notation	N	Crystal Structure	Cubic—face centered
Group	II A	Electronegativity	36.74
Category	Alkali earth metal	Electrical Conductance	0.218 micro-ohm
State	Solid	First Ionization Energy	141 kcal/gm-mole
Origin	Natural		
No. of Protons	20	Ionization Potential	6.1 ev
No. of Electrons	20	Heat of Vaporization	36.74 Kg-cal/gm-atom
Valence	+2		
Subshell Filling	s	Heat of Fusion	2.1 Kg-cal/gm-atom
Atomic Radius	1.97 Å	Specific Heat	0.149 cal/gm/°C
Covalent Radius	1.74 Å	Thermal Conductance	0.3 cal/cm²/cm/°C/sec
Ionic Radius	0.99 Å (+2)	Boiling Point	1440° C
Atomic Vol.	29.9 W/D	Melting Point	838° C
Core	Argon		

Ground State Electron Config.
$1s^2\ 2s^2\ 2p^6\ 3s^2\ 3p^6\ 3d^0\ 4s^2$

Radioactive Isotopes
Ca (at.wt.=41); ½life=1×10^5 years with decay via K.
Ca (at.wt.=45); ½life=160 days with decay via β^-.
Ca (at.wt.=47); ½life=4.7 days with decay via β^- and γ.
Radioactivity is induced.

Californium   Cf

Principal Quantum No.	7	Valence Electrons	$7s^2$
Atomic No.	98	Acid-Base Property	/
Atomic Wt.	251 △	Density	/
X-Ray Notation	Q	Crystal Structure	/
Group	III A		
Category	Actinon	Electro-negativity	/
State	Solid	Electrical Conductance	/
Origin	Synthetic		
No. of Protons	98	First Ioniza-tion Energy	/
No. of Electrons	98	Ionization Potential	/
Valence	+3	Heat of Vaporization	/
Subshell Filling	f	Heat of Fusion	/
Atomic Radius	/	Specific Heat	/
Covalent Radius	/	Thermal Conductance	/
Ionic Radius	/	Boiling Point	/
Atomic Vol.	/	Melting Point	/
Core	Krypton		

Ground State Electron Config.
$1s^2\ 2s^2\ 2p^6\ 3s^2\ 3p^6\ 3d^{10}$
$4s^2\ 4p^6\ 4d^{10}\ 4f^{14}\ 5s^2\ 5p^6$
$5d^{10}\ 5f^9\ 5g^0\ 6s^2\ 6p^6\ 6d^1$
$7s^2$

Radioactive Isotopes
Cf (at.wt.=246); ½life=35 hours with decay via α and γ; spontaneous fission.
Cf (at.wt.=249); ½life=500 years with decay via α and γ; spontaneous fission.
Cf (at.wt.=251); ½life=700 years with decay via γ.
Radioactivity is induced.

Carbon   C

Principal Quantum No.	2	Valence Electrons	$2s^2\ 2p^2$
Atomic No.	6	Acid-Base Property	Moderately acidic
Atomic Wt.	12.01115	Density	2.26 gm/ml
X-Ray Notation	L	Crystal Structure	Hexagonal
Group	IV A		
Category	Non-metal (semi-metal)	Electro-negativity	2.5
State	Solid	Electrical Conductance	0.0007 micro-ohm
Origin	Natural	First Ioniza-tion Energy	260 kcal/gm-mole
No. of Protons	6	Ionization Potential	11.3 ev
No. of Electrons	6	Heat of Vaporization	171.7 kg-cal/gm-atom
Valence	+4 (also, +2 and −4)	Heat of Fusion	/
Subshell Filling	p	Specific Heat	0.165 cal/gm/°C
Atomic Radius	0.914 Å	Thermal Conductance	0.057 cal/cm²/cm/°C/sec
Covalent Radius	0.77 Å	Boiling Point	4830° C
Ionic Radius	2.60 Å(−4); 0.15 Å(+4)	Melting Point	3726° C
Atomic Vol.	513 W/D		
Core	Neon		

Ground State Electron Config.
1s² 2s² 2p²

Radioactive Isotopes
C (at.wt.=14); ½life=5600 years with decay via β⁻.
Radioactive isotope occurs naturally.

Cerium — Ce

Principal Quantum No.	6	Valence Electrons	$4f^2\ 5d^0\ 6s^2$
Atomic No.	58	Acid-Base Property	Moderately basic
Atomic Wt.	140.12	Density	3.37 gm/ml
X-Ray Notation	P	Crystal Structure	Cubic—face centered
Group	V B		
Category	Lanthanon	Electro-negativity	1.1
State	Solid	Electrical Conductance	0.013 micro-ohm
Origin	Natural		
No. of Protons	58	First Ioniza-tion Energy	159 kcal/gm-mole
No. of Electrons	58	Ionization Potential	6.9 ev
Valence	+3, +4	Heat of Vaporization	95 kg-cal/gm-atom
Subshell Filling	f	Heat of Fusion	1.2 Kg-cal/gm-atom
Atomic Radius	1.81 Å	Specific Heat	0.042 cal/gm/°C
Covalent Radius	1.65 Å	Thermal Conductance	0.026 cal/cm²/cm/°C/sec
Ionic Radius	1.11 Å(+3); 1.01 Å(+4)	Boiling Point	3468° C
Atomic Vol.	21.0 W/D	Melting Point	795° C
Core	Krypton		

Ground State Electron Config.
$1s^2\ 2s^2\ 2p^6\ 3s^2\ 3p^6\ 3d^{10}$
$4s^2\ 4p^6\ 4d^{10}\ 4f^2\ 5s^2\ 5p^6$
$5d^0\ 5f^0\ 5g^0\ 6s^2$

Radioactive Isotopes
Ce (at.wt.=141); ½life=32 days with decay via β^- and γ.
Ce (at.wt.=143); ½life=33 hours with decay via β^- and γ.
Ce (at.wt.=144); ½life=285 days with decay via β^- and γ.
Radioactivity is induced.

Cesium   Cs

Principal Quantum No.	6	Valence Electrons	$6s^1$
Atomic No.	55	Acid-Base Property	Very basic
Atomic Wt.	132.905	Density	1.90 gm/ml
X-Ray Notation	P	Crystal Structure	Cubic—body centered
Group	I A	Electro-negativity	0.7
Category	Alkali earth metal	Electrical Conductance	0.053 micro-ohm
State	Liquid	First Ioniza-tion Energy	90 kcal/gm-mole
Origin	Natural	Ionization Potential	3.9 ev
No. of Protons	55	Heat of Vaporization	16.3 kg-cal/gm-atom
No. of Electrons	55	Heat of Fusion	0.50 Kg-cal/gm-atom
Valence	+1	Specific Heat	0.052 cal/gm/°C
Subshell Filling	s	Thermal Conductance	/
Atomic Radius	2.67 Å	Boiling Point	690° C
Covalent Radius	2.25 Å	Melting Point	28.8° C
Ionic Radius	1.69 Å (+2)		
Atomic Vol.	70 W/D		
Core	Krypton		

Ground State Electron Config.
$1s^2 \ 2s^2 \ 2p^6 \ 3s^2 \ 3p^6 \ 3d^{10}$
$4s^2 \ 4p^6 \ 4d^{10} \ 4f^0 \ 5s^2 \ 5p^6$
$5d^0 \ 5f^0 \ 5g^0 \ 6s^1$

Radioactive Isotopes
Cs (at.wt.=134); ½life=2.2 years with decay via β^- and γ.
Cs (at.wt.=135); ½life=3×10^6 years with decay via β^-.
Cs (at.wt.=137); ½life=30 years with decay via β^- and γ.
Radioactivity is induced.

Chlorine   Cl

Principal Quantum No.	3	Valence Electrons	$3p^5$
Atomic No.	17	Acid-Base Property	Very Acidic
Atomic Wt.	35.45273	Density	1.56 gm/ml
X-Ray Notation	M	Crystal Structure	Tetragonal
Group	VII A		
Category	Halogen	Electro-negativity	3.0
State	Gas	Electrical Conductance	/
Origin	Natural		
No. of Protons	17	First Ionization Energy	300 kcal/gm-mole
No. of Electrons	17	Ionization Potential	13.0 ev
Valence	+1, +3, +5, +7, −1 (also, +2, +4)	Heat of Vaporization	2.44 kg-cal/gm-atom
Subshell Filling	p	Heat of Fusion	0.77 kg-cal/gm-atom
		Specific Heat	0.116 cal/gm/°C
Atomic Radius	/	Thermal Conductance	0.00002 cal/cm²/cm/°C/sec
Covalent Radius	0.99 Å	Boiling Point	−34.7° C
Ionic Radius	1.81 Å(−1); 0.26 Å(+7)	Melting Point	−101° C
Atomic Vol.	18.7 W/D		
Core	Neon		

Ground State Electron Config.
$1s^2$ $2s^2$ $2p^6$ $3s^2$ $3p^5$

Radioactive Isotopes
Cl (at.wt.=36); ½life=3×10^5 years with decay via β^-.
Radioactivity is induced.

Chromium   Cr

Principal Quantum No.	4	Valence Electrons	$3d^5\ 4s^1\ (+6)$
Atomic No.	24	Acid-Base Property	Moderately acidic
Atomic Wt.	51.996	Density	7.19 gm/ml
X-Ray Notation	N	Crystal Structure	Cubic—body centered
Group	VI B		
Category	Heavy transitional metal	Electronegativity	1.6
State	Solid	Electrical Conductance	0.078 micro-ohm
Origin	Natural	First Ionization Energy	156 Kcal/gm-mole
No. of Protons	24	Ionization Potential	6.8 ev
No. of Electrons	24	Heat of Vaporization	72.97 Kg-cal/gm-atom
Valence	+2, +3, +6	Heat of Fusion	3.3 Kg-cal/gm-atom
Subshell Filling	d	Specific Heat	0.11 cal/gm/°C
Atomic Radius	1.27 Å	Thermal Conductance	0.16 cal/cm²/cm/°C/sec
Covalent Radius	/	Boiling Point	2665° C
Ionic Radius	0.69 Å(+3); 0.52 Å(+6)	Melting Point	1875° C
Atomic Vol.	7.23 W/D		
Core	Argon		

Ground State Electron Config.
1s² 2s² 2p⁶ 3s² 3p⁶ 3d⁵ 4s¹

Radioactive Isotopes
Cr (at.wt.=51); ½life=27 days with decay via γ and K. Radioactivity is induced.

Cobalt   Co

Principal Quantum No.	4	Valence Electrons	$3d^7\ 4s^2$
Atomic No.	27	Acid-Base Property	Amphoteric
Atomic Wt.	58.9332	Density	7.19 gm/ml
X-Ray Notation	N	Crystal Structure	Hexagonal
Group	VIII	Electronegativity	1.8
Category	Heavy transitional metal	Electrical Conductance	0.16 micro-ohm
State	Solid	First Ionization Energy	181 kcal/gm-mole
Origin	Natural	Ionization Potential	7.9 ev
No. of Protons	27	Heat of Vaporization	93 kg-cal/gm-atom
No. of Electrons	27	Heat of Fusion	3.64 kg-cal/gm-atom
Valence	+2, +3; (also, +4)	Specific Heat	0.099 cal/gm/°C
Subshell Filling	d	Thermal Conductance	0.16 cal/cm²/cm/°C/sec
Atomic Radius	1.25 Å	Boiling Point	2900° C
Covalent Radius	/	Melting Point	1495° C
Ionic Radius	0.78 Å(+2); 0.63 Å(+3)		
Atomic Vol.	6.7 W/D		
Core	Argon		

Ground State Electron Config.
1s² 2s² 2p⁶ 3s² 3p⁶ 3d⁷ 4s²

Radioactive Isotopes
Co (at.wt.=58); ½life=71 days
with decay via β^+, γ and K.
Co (at.wt.=60); ½life=5.27 years
with decay via β^- and γ.
Radioactivity is induced.

Copper   Cu (Cuprum)

Principal Quantum No.	4	Valence Electrons	$4s^1$ (+1)
Atomic No.	29	Acid-Base Property	Moderately basic
Atomic Wt.	63.54555	Density	8.96 gm/ml
X-Ray Notation	N	Crystal Structure	Cubic—face centered
Group	I B		
Category	Heavy transitional metal	Electronegativity	1.9
State	Solid	Electrical Conductance	0.593 micro-ohm
Origin	Natural	First Ionization Energy	178 kcal/gm-mole
No. of Protons	29	Ionization Potential	7.7 ev
No. of Electrons	29	Heat of Vaporization	72.8 Kg-cal/gm-atom
Valence	+1, +2; (also, +3)	Heat of Fusion	3.11 Kg-cal/gm-atom
Subshell Filling	d	Specific Heat	0.092 cal/gm/°C
Atomic Radius	1.28 Å	Thermal Conductance	0.94 cal/cm²/cm/°C/sec
Covalent Radius	1.38 Å	Boiling Point	2595° C
Ionic Radius	0.96 Å(+1); 0.69Å(+2)	Melting Point	1083° C
Atomic Vol.	7.1 W/D		
Core	Argon		

Ground State Electron Config.
$1s^2$ $2s^2$ $2p^6$ $3s^2$ $3p^6$ $3d^{10}$ $4s^1$

Radioactive Isotopes
Cu (at.wt.=64); ½life=12.8 hours with decay via β^-, β^+, γ and K. Radioactivity is induced.

Curium   Cm

Principal Quantum No.	7	Valence Electrons	6d¹ 7s² (+3)
Atomic No.	96	Acid-Base Property	/
Atomic Wt.	247 △	Density	/
X-Ray Notation	Q	Crystal Structure	/
Group	I B	Electro-negativity	/
Category	Actinon	Electrical Conductance	/
State	Solid		
Origin	Synthetic	Electrical Conductance	/
No. of Protons	96	First Ioniza-tion Energy	/
No. of Electrons	96	Ionization Potential	/
Valence	+3	Heat of Vaporization	/
Subshell Filling	f	Heat of Fusion	/
Atomic Radius	/	Specific Heat	/
Covalent Radius	/	Thermal Conductance	/
Ionic Radius	/	Boiling Point	/
Atomic Vol.	/	Melting Point	/
Core	Krypton		

Ground State Electron Config.
1s² 2s² 2p⁶ 3s² 3p⁶ 3d¹⁰
4s² 4p⁶ 4d¹⁰ 4f¹⁴ 5s² 5p⁶
5d¹⁰ 5f⁷ 5g⁰ 6s² 6p⁶ 6d¹
7s²

Radioactive Isotopes
Cm (at.wt.=243); ½life=35 years
with decay via α and γ.
Cm (at.wt.=245); ½life=11000 years
with decay via α and γ.
Cm (at.wt.=247); ½life=1 year
with decay via α and γ.
Radioactivity is induced.

Dysprosium   Dy

Principal Quantum No.	6		Valence Electrons	5d¹ 6s² (+3)
Atomic No.	66		Acid-Base Property	Slightly basic
Atomic Wt.	162.50		Density	8.54 gm/ml
X-Ray Notation	P		Crystal Structure	Hexagonal
Group	III A		Electro-negativity	/
Category	Lanthanons		Electrical	
State	Solid		Conductance	0.011 micro-ohm
Origin	Natural			
No. of Protons	66		First Ioniza-tion Energy	157 kcal/gm-mole
No. of Electrons	66		Ionization Potential	6.8 ev
Valence	+3		Heat of Vaporization	67 Kg-cal/gm-atom
Subshell Filling	f		Heat of Fusion	4.1 Kg-cal/gm-atom
Atomic Radius	1.77 Å		Specific Heat	0.041 cal/gm/°C
Covalent Radius	1.59 Å		Thermal Conductance	0.024 cal/cm²/cm/°C/sec
Ionic Radius	0.99 Å(+3)		Boiling Point	2600° C
Atomic Vol.	19.0 W/D		Melting Point	1407° C
Core	Krypton			

Ground State Electron Config.
1s² 2s² 2p⁶ 3s² 3p⁶ 3d¹⁰
4s² 4p⁶ 4d¹⁰ 4f⁹ 5s² 5p⁶
5d¹ 5f⁰ 5g⁰ 6s²

Radioactive Isotopes

None

Einsteinium   Es

Principal Quantum No.	7	Valence Electrons	$6d^1\ 7s^2$	$(+3)$
Atomic No.	99	Acid-Base Property	/	
Atomic Wt.	254 △	Density	/	
X-Ray Notation	Q	Crystal Structure	/	
Group	IV A	Electro-negativity	/	
Category	Actinons			
State	Solid	Electrical Conductance	/	
Origin	Synthetic			
No. of Protons	99	First Ioniza-tion Energy	/	
No. of Electrons	99	Ionization Potential	/	
Valence	+3	Heat of Vaporization	/	
Subshell Filling	f	Heat of Fusion	/	
Atomic Radius	/	Specific Heat	/	
Covalent Radius	/	Thermal Conductance	/	
Ionic Radius	/	Boiling Point	/	
Atomic Vol.	/	Melting Point	/	
Core	Krypton			

Ground State Electron Config.
$1s^2\ 2s^2\ 2p^6\ 3s^2\ 3p^6\ 3d^{10}$
$4s^2\ 4p^6\ 4d^{10}\ 4f^{14}\ 5s^2\ 5p^6$
$5d^{10}\ 5f^{10}\ 5g^0$
$6s^{(2}\ 6p^6\ 6d^1\ 7s^{2)}$

Radioactive Isotopes
Es (at.wt.=253); ½life=20 days
with decay via α and γ; spontaneous fission.
Es (at.wt.=254); ½life=1 year
with decay via α; spontaneous fission.
Radioactivity is induced.

38

Erbium   Er

Principal Quantum No.	6	Valence Electrons	5d¹ 6s² (+3)
Atomic No.	68	Acid-Base Property	Slightly basic
Atomic Wt.	167.26	Density	9.05 gm/ml
X-Ray Notation	P	Crystal Structure	Hexagonal
Group	V A	Electronegativity	1.2
Category	Lanthanons		
State	Solid	Electrical Conductance	0.012 micro-ohm
Origin	Natural		
No. of Protons	68	First Ionization Energy	/
No. of Electrons	68	Ionization Potential	/
Valence	+3	Heat of Vaporization	67 Kg-cal/gm-atom
Subshell Filling	f	Heat of Fusion	4.1 Kg-cal/gm-atom
Atomic Radius	1.75 Å	Specific Heat	0.040 cal/gm/°C
Covalent Radius	1.57 Å	Thermal Conductance	0.023 cal/cm²/cm/°C/sec
Ionic Radius	0.96 Å(+3)	Boiling Point	2900° C
Atomic Vol.	18.4 W/D	Melting Point	1497° C
Core	Krypton		

Ground State Electron Config.
1s² 2s² 2p⁶ 3s² 3p⁶ 3d¹⁰
4s² 4p⁶ 4d¹⁰ 4f¹¹ 5s² 5p⁶
5d¹ 5f⁰ 5g⁰ 6s²

Radioactive Isotopes
None

Europium   Eu

Principal Quantum No.	6		Valence Electrons	$6s^2$ (+2)
Atomic No.	63		Acid-Base Property	Slightly basic
Atomic Wt.	151.96		Density	5.26 gm/ml
X-Ray Notation	P		Crystal Structure	Cubic—face centered
Group	VIII B		Electronegativity	/
Category	Lanthanon		Electrical Conductance	0.012 micro-ohm
State	Solid			
Origin	Natural			
No. of Protons	63		First Ionization Energy	131 kcal/gm-mole
No. of Electrons	63		Ionization Potential	5.7 ev
Valence	+3 (also, +2)		Heat of Vaporization	42 Kg-cal/gm-atom
Subshell Filling	f		Heat of Fusion	2.2 Kg-cal/gm-atom
Atomic Radius	2.04 Å		Specific Heat	0.039 cal/gm/°C
Covalent Radius	1.85 Å		Thermal Conductance	/
Ionic Radius	1.12 Å(+2)		Boiling Point	1439° C
Atomic Vol.	28.9 W/D		Melting Point	826° C
Core	Krypton			

Ground State Electron Config.
$1s^2$ $2s^2$ $2p^6$ $3s^2$ $3p^6$ $3d^{10}$
$4s^2$ $4p^6$ $4d^{10}$ $4f^7$ $5s^2$ $5p^6$
$5d^0$ $5f^0$ $5g^0$ $6s^2$

Radioactive Isotopes
Eu (at.wt.=154); ½life=16 years
wtih decay via β^- and γ.
Eu (at.wt.=155); ½life=1.7 years
with decay via β^- and γ.
Radioactivity is induced.

Fermium   Fm

Principal Quantum No.	7		Valence Electrons	$6d^1\ 7s^2\ (+3)$
Atomic No.	100		Acid-Base Property	/
Atomic Wt.	253 △		Density	/
X-Ray Notation	Q		Crystal	/
Group	V A		Structure	/
Category	Actinon		Electronegativity	/
State	Solid			
Origin	Synthetic		Electrical Conductance	/
No. of Protons	100		First Ionization Energy	/
No. of Electrons	100		Ionization Potential	/
Valence	+3		Heat of Vaporization	/
Subshell Filling	f		Heat of Fusion	/
Atomic Radius	/		Specific Heat	/
Covalent Radius	/		Thermal Conductance	/
Ionic Radius	/		Boiling Point	/
Atomic Vol.	/		Melting Point	
Core	Krypton			

Ground State Electron Config.
$1s^2\ 2s^2\ 2p^6\ 3s^2\ 3p^6\ 3d^{10}$
$4s^2\ 4p^6\ 4d^{10}\ 4f^{14}\ 5s^2\ 5p^6$
$5d^{10}\ 5f^{11}\ 5g^0$
$6s^{(2}\ 6p^6\ 6d^1\ 7s^2)$

Radioactive Isotopes
Fm (at.wt.=255); ½life=20 hours
with decay via α.
Radioactivity is induced.

Fluorine   F

Principal Quantum No.	2	Valence Electrons	$2p^5$ (-1)
Atomic No.	9	Acid-Base Property	Very acidic
Atomic Wt.	18.9984	Density	1.56 gm/ml
X-Ray Notation	L	Crystal Structure	Tetragonal
Group	VII A	Electro-negativity	3.0
Category	Halogen		
State	Gas	Electrical Conductance	/
Origin	Natural		
No. of Protons	9	First Ioniza-tion Energy	300 kcal/gm-mole
No. of Electrons	9	Ionization Potential	17.4 ev
Valence	-1 (also, $+1$)	Heat of Vaporization	2.44 Kg-cal/gm-atom
Subshell Filling	p	Heat of Fusion	0.77 Kg-cal/gm-atom
Atomic Radius	/	Specific Heat	0.116 cal/gm/°C
Covalent Radius	0.99 Å	Thermal Conductance	0.00002 cal/cm²/cm/°C/sec
Ionic Radius	1.81 Å(-1); 0.26 Å$(+1)$	Boiling Point	$-188.2°$ C
Atomic Vol.	18.7 W/D	Melting Point	$-219.6°$ C
Core	/		

Ground State Electron Config.
$1s^2$ $2s^2$ $2p^5$

Radioactive Isotopes
None.

Francium		Fr	
Principal Quantum No.	7	Valence Electrons	$7s^1$ (+1)
Atomic No.	87	Acid-Base Property	Very basic
Atomic Wt.	223 △		
X-Ray Notation	Q	Density	/
Group	I A	Crystal Structure	Cubic—body centered
Category	Alkali earth metal	Electro-negativity	0.7
State	Solid	Electrical Conductance	/
Origin	Natural	First Ioniza-tion Energy	/
No. of Protons	87		
No. of Electrons	87	Ionization Potential	/
Valence	+1	Heat of Vaporization	/
Subshell Filling	s	Heat of Fusion	/
Atomic Radius	/	Specific Heat	/
Covalent Radius	/	Thermal Conductance	/
Ionic Radius	1.76 Å(+1)	Boiling Point	/
Atomic Vol.	/	Melting Point	27° C
Core	Krypton		

Ground State Electron Config.
$1s^2$ $2s^2$ $2p^6$ $3s^2$ $3p^6$ $3d^{10}$
$4s^2$ $4p^6$ $4d^{10}$ $4f^{14}$ $5s^2$ $5p^6$
$5d^{10}$ $5f^0$ $5g^0$ $6s^2$ $6p^6$ $6d^0$
$7s^1$

Radioactive Isotopes
Fr (at.wt.=223); ½life=22 minutes
with decay via α, β^- and γ.
Radioactive isotope occurs naturally.

Gadolinium   Gd

Principal Quantum No.	6	Valence Electrons	5d¹ 6s² (+3)
Atomic No.	64	Acid-Base Property	Moderately basic
Atomic Wt.	157.25	Density	7.89 gm/ml
X-Ray Notation	P	Crystal Structure	Hexagonal
Group	I B		
Category	Lanthanon	Electro-negativity	1.1
State	Solid	Electrical Conductance	0.007 micro-ohm
Origin	Natural		
No. of Protons	64	First Ioniza-tion Energy	142 kcal/gm-mole
No. of Electrons	64	Ionization Potential	6.2 ev
Valence	+3	Heat of Vaporization	72 Kg-cal/gm-atom
Subshell Filling	f	Heat of Fusion	3.70 Kg-cal/gm-atom
Atomic Radius	1.79 Å	Specific Heat	0.071 cal/gm/°C
Covalent Radius	1.61 Å	Thermal Conductance	0.021 cal/cm²/cm/°C/sec
Ionic Radius	1.02 Å(+3)	Boiling Point	3000° C
Atomic Vol.	19.9 W/D	Melting Point	1312° C
Core	Krypton		

Ground State Electron Config.
1s² 2s² 2p⁶ 3s² 3p⁶ 3d¹⁰
4s² 4p⁶ 4d¹⁰ 4f⁷ 5s² 5p⁶
5d¹ 5f⁰ 5g⁰ 6s²

Radioactive Isotopes
Gd (at.wt.=153); ½life=236 days
  with decay via γ, e⁻ and K.
Gd (at.wt.=159); ½life=18 hours
  with decay via β⁻ and γ.
Radioactivity is induced.

Gallium   Ga

Principal Quantum No.	4	Valence Electrons	$4p^1$ (+3)
Atomic No.	31	Acid-Base Property	Amphoteric
Atomic Wt.	69.72	Density	5.91 gm/ml
X-Ray Notation	N	Crystal Structure	Orthorhombic
Group	III A	Electro-negativity	1.6
Category	Metal		
State	Liquid	Electrical Conductance	0.58 micro-ohm
Origin	Natural		
No. of Protons	31	First Ioniza-tion Energy	1.38 kcal/gm-mole
No. of Electrons	31	Ionization Potential	6.0 ev
Valence	+3 (also, +2)	Heat of Vaporization	/
Subshell Filling	p	Heat of Fusion	1.34 kg-cal/gm-atom
Atomic Radius	1.41 Å	Specific Heat	0.079 cal/gm/°C
Covalent Radius	1.26 Å	Thermal Conductance	/
Ionic Radius	1.48 Å(+2); 0.62 Å(+3)	Boiling Point	2237° C
		Melting Point	29.8° C
Atomic Vol.	11.8 W/D		
Core	Argon		

Ground State Electron Config.
$1s^2\ 2s^2\ 2p^6\ 3s^2\ 3p^6\ 3d^{10}$
$4s^2\ 4p^1$

Radioactive Isotopes
Ga (at.wt.=72); ½life=14.11 hours
with decay via β^- and γ.
Radioactivity is induced.

Germanium   Ge

Principal Quantum No.	4	Valence Electrons	$4s^2 \ 4p^2 \ (+2)$
Atomic No.	32	Acid-Base Property	Amphoteric
Atomic Wt.	72.59	Density	5.32 gm/ml
X-Ray Notation	N	Crystal Structure	Diamond
Group	IV A		
Category	Metal	Electro-negativity	1.8
State	Solid	Electrical Conductance	0.022 micro-ohm
Origin	Natural		
No. of Protons	32	First Ioniza-ation Energy	187 kcal/gm-mole
No. of Electrons	32	Ionization Potential	8.1 ev
Valence	+2, +4, −4	Heat of Vaporization	68 kg-cal/gm-atom
Subshell Filling	p	Heat of Fusion	7.6 kg-cal/gm-atom
Atomic Radius	1.37 Å	Specific Heat	0.073 cal/gm/°C
Covalent Radius	1.22 Å	Thermal Conductance	0.14 cal/cm²/cm/°C/sec
Ionic Radius	0.93 Å(+2); 0.53 Å(+4)	Boiling Point	2830° C
Atomic Vol.	13.6 W/D	Melting Point	937.4° C
Core	Argon		

Ground State Electron Config.
$1s^2 \ 2s^2 \ 2p^6 \ 3s^2 \ 3p^6 \ 3d^{10}$
$4s^2 \ 4p^2$

Radioactive Isotopes
Ge (at.wt.=71); ½life=11 days
with decay via K.
Radioactivity is induced.

Gold   Au (Aurum)

Principal Quantum No.	6	Valence Electrons	6s^1 (+1)
Atomic No.	79	Acid-Base Property	Amphoteric
Atomic Wt.	196.967	Density	19.3 gm/ml
X-Ray Notation	P	Crystal Structure	Cubic—face centered
Group	I B	Electronegativity	2.4
Category	Heavy transitional metal	Electrical Conductance	0.42 micro-ohm
State	Solid	First Ionization Energy	213 kcal/gm-mole
Origin	Natural		
No. of Protons	79	Ionization Potential	9.2 ev
No. of Electrons	79	Heat of Vaporization	81.8 Kg-cal/gm-atom
Valence	+1, +3 (also, +2)	Heat of Fusion	3.03 Kg-cal/gm-atom
Subshell Filling	s	Specific Heat	0.031 cal/gm/°C
Atomic Radius	1.44 Å	Thermal Conductance	0.02 cal/cm^2/cm/°C/sec
Covalent Radius	1.50 Å	Boiling Point	2970° C
Ionic Radius	1.37 Å(+1)	Melting Point	1063° C
Atomic Vol.	10.2 W/D		
Core	Krypton		

Ground State Electron Config.
1s^2 2s^2 2p^6 3s^2 3p^6 3d^{10}
4s^2 4p^6 4d^{10} 4f^{14} 5s^2 5p^6
5d^{10} 5f^0 5g^0 6s^1

Radioactive Isotopes
Au (at.wt.=198); ½life=2.7 days
with decay via β^- and γ.
Radioactivity is induced.

Hafnium   Hf

Quantum No.		Valence	
Principal	6	Electrons	$5d^2\ 6s^2\ (+4)$
Atomic No.	72	Acid-Base	
Atomic Wt.	178.49	Property	Amphoteric
X-Ray		Density	13.1 gm/ml
Notation	P	Crystal	
Group	IV B	Structure	Hexagonal
Category	Heavy transi-	Electro-	
	tional metal	negativity	1.3
State	Solid	Electrical	
		Conductance	0.031 micro-ohm
Origin	Natural	First Ioniza-	
No. of		tion Energy	127 Kcal/gm-mole
Protons	72	Ionization	
No. of		Potential	5.5 ev
Electrons	72	Heat of	
Valence	+4	Vaporization	155 Kg-cal/gm-atom
Subshell		Heat of	
Filling	d	Fusion	5.2 Kg-cal/gm-atom
Atomic		Specific Heat	0.035 cal/gm/°C
Radius	1.58 Å	Thermal	0.22
Covalent		Conductance	cal/cm²/cm/°C/sec
Radius	/	Boiling Point	5400° C
Ionic Radius	0.81 Å	Melting Point	2222° C
Atomic Vol.	13.6 W/D		
Core	Krypton		

Ground State Electron Config.
$1s^2\ 2s^2\ 2p^6\ 3s^2\ 3p^6\ 3d^{10}$
$4s^2\ 4p^6\ 4d^{10}\ 4f^{14}\ 5s^2\ 5p^6$
$5d^2\ 5f^0\ 5g^0\ 6s^2$

Radioactive Isotopes
Hf (at.wt.=181); ½life=45 days
with decay via β^-, γ and e^-.
Radioactivity is induced.

Helium   He

Principal Quantum No.	1	Valence Electrons	0
Atomic No.	2	Acid-Base Property	/
Atomic Wt.	4.0026	Density	0.126 gm/ml
X-Ray Notation	K	Crystal Structure	Hexagonal
Group	Inert gas		
Category	Inert gas	Electro-negativity	/
State	Gas	Electrical Conductance	/
Origin	Natural		
No. of Protons	1	First Ionization Energy	567 Kcal/gm-mole
No. of Electrons	1	Ionization Potential	24.6 ev
Valence	0	Heat of Vaporization	0.020 Kg-cal/gm-atom
Subshell Filling	s	Heat of Fusion	0.005 Kg-cal/gm-atom
Atomic Radius	/	Specific Heat	1.25 cal/gm/°C
Covalent Radius	0.93 Å	Thermal Conductance	0.0003 cal/cm²/cm/°C/sec
Ionic Radius	/	Boiling Point	−268.9° C
Atomic Vol.	31.8 W/D	Melting Point	−269.7° C
Core	/		

Ground State Electron Config.
$1s^2$

Radioactive Isotopes
None

49

Holmium   Ho

Principal Quantum No.	6		Valence Electrons	5d¹ 6s² (+3)
Atomic No.	67		Acid-Base Property	Slightly basic
Atomic Wt.	164.930		Density	8.80 gm/ml
X-Ray Notation	P		Crystal Structure	Hexagonal
Group	IV A		Electro-negativity	1.2
Category	Lanthanon			
State	Solid		Electrical Conductance	0.011 micro-ohm
Origin	Natural			
No. of Protons	67		First Ioniza-tion Energy	/
No. of Electrons	67		Ionization Potential	/
Valence	+3		Heat of Vaporization	67 Kg-cal/gm-atom
Subshell Filling	f		Heat of Fusion	4.1 Kg-cal/gm-atom
Atomic Radius	1.76 Å		Specific Heat	0.039 cal/gm/°C
Covalent Radius	1.58 Å		Thermal Conductance	/
Ionic Radius	0.97 Å		Boiling Point	2600° C
Atomic Vol.	18.7 W/D		Melting Point	1461° C
Core	Krypton			

Ground State Electron Config.
1s² 2s² 2p⁶ 3s² 3p⁶ 4d¹⁰
4s² 4p⁶ 4d¹⁰ 4f¹⁰ 5s² 5p⁶
5d¹ 5f⁰ 5g⁰ 6s²

Radioactive Isotopes
Ho (at.wt.=166); ½life=27.3 hours
with decay via β⁻ and γ.
Radioactivity is induced.

Hydrogen   H

Principal Quantum No.	1		Valence Electrons	$1s^1$
Atomic No.	1		Acid-Base Property	Amphoteric
Atomic Wt.	1.00797		Density	0.071 gm/ml
X-Ray Notation	K		Crystal Structure	Hexagonal
Group	I A		Electro-negativity	2.1
Category	Gas		Electrical Conductance	/
State	Gas			
Origin	Natural			
No. of Protons	1		First Ioniza-tion Energy	313 kcal/gm-mole
No. of Electrons	1		Ionization Potential	13.6 ev
Valence	+1 (also, −1)		Heat of Vaporization	0.108 Kg-cal/gm-atom
Subshell Filling	s		Heat of Fusion	0.014 Kg-cal/gm-atom
Atomic Radius	/		Specific Heat	3.45 cal/gm/°C
Covalent Radius	0.37 Å		Thermal Conductance	0.0004 cal/cm²/cm/°C/sec
Ionic Radius	2.08 Å(−1)		Boiling Point	−252.7° C
Atomic Vol.	14.1 W/D		Melting Point	−259.2° C
Core	/			

Ground State Electron Config.
$1s^1$

Radioactive Isotopes
H (at.wt.=3); ½life=12.3 years
with decay via β^-.
Radioactive isotope occurs naturally.

Indium   In

Principal Quantum No.	5	Valence Electrons: $5s^2\ 6p^1$ (+3)
Atomic No.	49	Acid-Base Property: Amphoteric
Atomic Wt.	114.82	Density: 7.31 gm/ml
X-Ray Notation	Q	Crystal Structure: Tetragonal
Group	III A	Electronegativity: 1.7
Category	Heavy transitional metal	Electrical Conductance: 0.111 micro-ohm
State	Solid	First Ionization Energy: 1.33 Kcal/gm-mole
Origin	Natural	
No. of Protons	49	Ionization Potential: 5.8 ev
No. of Electrons	49	Heat of Vaporization: 53.7 Kg-cal/gm-atom
Valence	+3 (also, +1, +2)	Heat of Fusion: 0.78 Kg-cal/gm-atom
Subshell Filling	p	Specific Heat: 0.057 cal/gm/°C
Atomic Radius	1.66 Å	Thermal Conductance: 0.057 cal/cm²/cm/°C/sec
Covalent Radius	1.44 Å	Boiling Point: 2000° C
Ionic Radius	1.32 Å(+1); 0.81 Å(+3)	Melting Point: 156.2° C
Atomic Vol.	15.7 W/D	
Core	Krypton	

Ground State Electron Config.
$1s^2\ 2s^2\ 2p^6\ 3s^2\ 3p^6\ 3d^{10}$
$4s^2\ 4p^6\ 4d^{10}\ 4f^0\ 5s^2\ 5p^1$

Radioactive Isotopes
In (at.wt.=114); ½life=50 days
with decay via γ.
Radioactivity is induced.

Iodine   I

Principal Quantum No.	5	Valence Electrons	$5s^2\ 5p^5\ (+7)$
Atomic No.	53	Acid-Base Property	Very acidic
Atomic Wt.	126.9044	Density	4.94 gm/ml
X-Ray Notation	Q	Crystal Structure	Orthorhombic
Group	VII A	Electro-negativity	2.5
Category	Halogen		
State	Solid	Electrical Conductance	10^{-15} micro-ohm
Origin	Natural		
No. of Protons	53	First Ioniza-tion Energy	241 Kcal/gm-mole
No. of Electrons	53	Ionization Potential	10.4 ev
Valence	+1, +5, +7 (also, +3, +4 and −1)	Heat of Vaporization	5.2 Kg-cal/gm-atom
Subshell Filling	p	Heat of Fusion	1.87 Kg-cal/gm-atom
		Specific Heat	0.052 cal/gm/°C
Atomic Radius	/	Thermal Conductance	0.001 cal/cm²/cm/°C/sec
Covalent Radius	1.33 Å	Boiling Point	183° C
Ionic Radius	2.16 Å(−1); 0.50 Å(+7)	Melting Point	113.7° C
Atomic Vol.	25.7 W/D		
Core	Krypton		

Ground State Electron Config.
1s² 2s² 2p⁶ 3s² 3p⁶ 3d¹⁰
4s² 4p⁶ 4d¹⁰ 4f⁰ 5s² 5p⁵

Radioactive Isotopes
I (at.wt.=129); ½life=10⁷ years
  with decay via β⁺, γ and e⁻.
I (at.wt.=131); ½life=8.05 days
  with decay via β⁻ and γ.
Radioactivity is induced.

Iridium   Ir

Principal Quantum No.	6	Valence Electrons	5d^7 6s^2
Atomic No.	77	Acid-Base Property	Moderately basic
Atomic Wt.	192.2	Density	22.5 gm/ml
X-Ray Notation	P	Crystal Structure	Cubic—face centered
Group	VIII	Electronegativity	2.2
Category	Transitional heavy metal	Electrical Conductance	0.189 micro-ohm
State	Solid	First Ionization Energy	212 Kcal/gm-mole
Origin	Natural		
No. of Protons	77	Ionization Potential	9.2 ev
No. of Electrons	77	Heat of Vaporization	152 Kg-cal/gm-atom
Valence	+3, +4, +6 (also, +1, +2)	Heat of Fusion	6.6 Kg-cal/gm-atom
Subshell Filling	d	Specific Heat	0.031 cal/gm/°C
Atomic Radius	1.36 Å	Thermal Conductance	0.14 cal/cm^2/cm/°C/sec
Covalent Radius	/	Boiling Point	5300° C
Ionic Radius	0.66 Å(+4)	Melting Point	2454° C
Atomic Vol.	8.54 W/D		
Core	Krypton		

Ground State Electron Config.
1s^2 2s^2 2p^6 3s^2 3p^6 3d^{10}
4s^2 4p^6 4d^{10} 4f^{14} 5s^2 5p^6
5d^7 5f^0 5g^0 6s^2

Radioactive Isotopes
Ir (at.wt.=192); ½life=74.4 days
with decay via β^- and γ.
Radioactivity is induced.

Iron   Fe (Ferrum)

Principal Quantum No.	4	Valence Electrons	$3d^6$ $4s^2$
Atomic No.	26	Acid-Base Property	Amphoteric
Atomic Wt.	55.847	Density	7.86 gm/ml
X-Ray Notation	N	Crystal Structure	Cubic—body centered
Group	VIII	Electro-negativity	1.8
Category	Heavy transitional metal	Electrical Conductance	0.10 micro-ohm
State	Solid	First Ionization Energy	182 Kcal/gm-mole
Origin	Natural	Ionization Potential	7.9 ev
No. of Protons	26	Heat of Vaporization	84.6 Kg-cal/gm-atom
No. of Electrons	26	Heat of Fusion	3.67 Kg-cal/gm-atom
Valence	+2, +3 (also, +4, +6)	Specific Heat	0.11 cal/gm/°C
Subshell Filling	d	Thermal Conductance	0.18 cal/cm²/cm/°C/sec
Atomic Radius	1.26 Å	Boiling Point	3000° C
Covalent Radius	/	Melting Point	1536° C
Ionic Radius	0.76 Å(+2); 0.64 Å(+3)		
Atomic Vol.	7.1 W/D		
Core	Argon		

Ground State Electron Config.
$1s^2$ $2s^2$ $2p^6$ $3s^2$ $3p^6$ $3d^6$ $4s^2$

Radioactive Isotopes
Fe (at.wt.=55); ½life=2.9 years
with decay via K.
Fe (at.wt.=59); ½life=45 days
with decay via β^- and γ.
Radioactivity is induced.

Krypton   Kr

Principal Quantum No.	4		Valence Electrons	0
Atomic No.	36		Acid-Base Property	/
Atomic Wt.	83.80			
X-Ray Notation	N		Density	2.6 gm/ml
Group	Inert gas		Crystal Structure	Cubic—face centered
Category	Gas		Electro-negativity	/
State	Gas		Electrical Conductance	/
Origin	Natural			
No. of Protons	36		First Ioniza-tion Energy	323 Kcal/gm-mole
No. of Electrons	36		Ionization Potential	14.0 ev
Valence	0		Heat of Vaporization	2.16 Kg-cal/gm-atom
Subshell Filling	p		Heat of Fusion	0.39 Kg-cal/gm-atom
Atomic Radius	/		Specific Heat	/
Covalent Radius	1.89 Å		Thermal Conductance	0.00002 cal/cm^2/cm/°C/sec
Ionic Radius	/		Boiling Point	−152° C
Atomic Vol.	32.2 W/D		Melting Point	−157° C
Core	/			

Ground State Electron Config.
1s^2 2s^2 2p^6 3s^2 3p^6 3d^{10}
4s^2 4p^6

Radioactive Isotopes
None.

Lanthanum   La

Principal Quantum No.	6		Valence Electrons	5d¹ 6s² (+3)
Atomic No.	57		Acid-Base Property	Basic
Atomic Wt.	138.91		Density	6.17 gm/ml
X-Ray Notation	P		Crystal Structure	Hexagonal
Group	IV B		Electro-negativity	1.1
Category	Lanthanon			
State	Solid		Electrical Conductance	0.017 micro-ohm
Origin	Natural			
No. of Protons	57		First Ioniza-tion Energy	129 Kcal/gm-mole
No. of Electrons	57		Ionization Potential	5.6 ev
Valence	+3		Heat of Vaporization	96 Kg-cal/gm-atom
Subshell Filling	f		Heat of Fusion	1.5 Kg-cal/gm-atom
Atomic Radius	1.87 Å		Specific Heat	0.045 cal/gm/°C
Covalent Radius	1.69 Å		Thermal Conductance	0.033 cal/cm²/cm/°C/sec
Ionic Radius	1.15 Å(+3)		Boiling Point	3470° C
Atomic Vol.	22.5 W/D		Melting Point	920° C
Core	Krypton			

Ground State Electron Config.
1s² 2s² 2p⁶ 3s² 3p⁶ 3d¹⁰
4s² 4p⁶ 4d¹⁰ 4f⁰ 5s² 5p⁶
5d¹ 5f⁰ 5g⁰ 6s²

Radioactive Isotopes
La (at.wt.=140); ½life=40.2 hours
with decay via β⁻ and γ.
Radioactivity is induced.

Lawrencium*   Lw*

Principal Quantum No.	7	Valence Electrons	6d¹ 7s² (+3)
Atomic No.	103	Acid-Base Property	/
Atomic Wt.	257 △	Density	/
X-Ray Notation	Q	Crystal Structure	/
Group	VIII A	Electro-negativity	/
Category	Actinons	Electrical Conductance	/
State	Solid		
Origin	Synthetic	First Ioniza-tion Energy	/
No. of Protons	103	Ionization Potential	/
No. of Electrons	103	Heat of Vaporization	/
Valence	+3 (?)	Heat of Fusion	/
Subshell Filling	f	Specific Heat	/
Atomic Radius	/	Thermal Conductance	/
Covalent Radius	/	Boiling Point	/
Ionic Radius	/	Melting Point	/
Atomic Vol.	/		
Core	Krypton		

Ground State Electron Config.
1s² 2s² 2p⁶ 3s² 3p⁶ 3d¹⁰
4s² 4p⁶ 4d¹⁰ 4f¹⁴ 5s² 5p⁶
5d¹⁰ 5f¹⁴ 5g⁰
6s⁽² 6p⁶ 6d¹ 7s²⁾

Radioactive Isotopes
Lw (at.wt.=257); ½life=
with decay via
Radioactivity is induced.

* Proposed element; not accepted officially.

Lead		Pb (Plumbum)	
Principal Quantum No.	6	Valence Electrons	$6s^2$ $6p^2$ $(+4)$
Atomic No.	82	Acid-Base Property	Amphoteric
Atomic Wt.	207.19	Density	11.4 gm/ml
X-Ray Notation	P	Crystal Structure	Cubic—face centered
Group	IV A	Electro-negativity	1.8
Category	Heavy transitional metal	Electrical Conductance	0.046 micro-ohm
State	Solid	First Ionization Energy	171 Kcal/gm-mole
Origin	Natural		
No. of Protons	82	Ionization Potential	7.4 ev
No. of Electrons	82	Heat of Vaporization	42.4 Kg-cal/gm-atom
Valence	+4 (also, +2)		
Subshell Filling	p	Heat of Fusion	1.22 Kg-cal/gm-atom
Atomic Radius	1.75 Å	Specific Heat	0.031 cal/gm/°C
Covalent Radius	1.47 Å	Thermal Conductance	0.083 cal/cm²/cm/°C/sec
Ionic Radius	1.20 Å(+2); 0.84 Å(+2)	Boiling Point	1725° C
		Melting Point	327.4° C
Atomic Vol.	18.3 W/D		
Core	Krypton		

Ground State Electron Config.
1s² 2s² 2p⁶ 3s² 3p⁶ 3d¹⁰
4s² 4p⁶ 4d¹⁰ 4f¹⁴ 5s² 5p⁶
5d¹⁰ 5f⁰ 5g⁰ 6s² 6p²

Radioactive Isotopes
Pb (at.wt.=210); ½life=19.4 years with decay via β⁻, γ and e⁻. This isotope occurs naturally.
Pb (at.wt.=202); ½life=10⁵ years with decay via L. Radioactivity is induced.

Lithium   Li

Principal Quantum No.	2	Valence Electrons	$2s^1$ (+1)
Atomic No.	3	Acid-Base Property	Very basic
Atomic Wt.	6.939	Density	0.53 gm/ml
X-Ray Notation	L	Crystal Structure	Cubic—body centered
Group	I A		
Category	Alkali earth metal	Electro-negativity	1.0
State	Solid	Electrical Conductance	0.108 micro-ohm
Origin	Natural	First Ioniza-tion Energy	124 Kcal/gm-mole
No. of Protons	3	Ionization Potential	5.4 ev
No. of Electrons	3	Heat of Vaporization	32.48 Kg-cal/gm-atom
Valence	+1		
Subshell Filling	s	Heat of Fusion	0.72 Kg-cal/gm-atom
Atomic Radius	1.55 Å	Specific Heat	0.79 cal/gm/°C
Covalent Radius	1.34 Å	Thermal Conductance	0.17 cal/cm²/cm/°C/sec
Ionic Radius	0.60 Å(+1)	Boiling Point	1330° C
Atomic Vol.	13.1 W/D	Melting Point	108.5° C
Core	/		

Ground State Electron Config.
$1s^2$ $2s^1$

Radioactive Isotopes
None.

Lutetium   Lu

Principal Quantum No.	6		Valence Electrons	$5d^1$ $6s^2$ $(+3)$
Atomic No.	71		Acid-Base Property	Slightly basic
Atomic Wt.	174.97		Density	9.84 gm/ml
X-Ray Notation	P		Crystal Structure	Hexagonal
Group	VIII A			
Category	Lanthanon		Electro-negativity	1.2
State	Solid		Electrical Conductance	0.015 micro-ohm
Origin	Natural			
No. of Protons	71		First Ionization Energy	115 Kcal/gm-mole
No. of Electrons	71		Ionization Potential	5.0 ev
Valence	+3		Heat of Vaporization	90 Kg-cal/gm-atom
Subshell Filling	f		Heat of Fusion	4.6 Kg-cal/gm-atom
Atomic Radius	1.74 Å		Specific Heat	0.037 cal/gm/°C
Covalent Radius	1.56 Å		Thermal Conductance	/
Ionic Radius	0.93 Å		Boiling Point	3327° C
Atomic Vol.	17.8 W/D		Melting Point	1625° C
Core	Krypton			

Ground State Electron Config.
$1s^2$ $2s^2$ $2p^6$ $3s^2$ $3p^6$ $3d^{10}$
$4s^2$ $4p^6$ $4d^{10}$ $4f^{14}$ $5s^2$ $5p^6$
$5d^1$ $5f^0$ $5g^0$ $6s^2$

Radioactive Isotopes
Lu (at.wt.=176); ½life=10^{10} years
with decay via β^-, γ and K. This
isotope occurs naturally.
Lu (at.wt.=177); ½life=6.8 days
with decay via β^- and γ. Radioactivity is induced.

Magnesium   Mg

Principal Quantum No.	3	Valence Electrons	$3s^2$ (+2)
Atomic No.	12	Acid-Base Property	Moderately basic
Atomic Wt.	24.312	Density	0.97 gm/ml
X-Ray Notation	M	Crystal Structure	Hexagonal
Group	II A	Electronegativity	1.2
Category	Alkali earth metal	Electrical Conductance	0.224 micro-ohm
State	Solid	First Ionization Energy	176 Kcal/gm-mole
Origin	Natural		
No. of Protons	12	Ionization Potential	7.6 ev
No. of Electrons	12	Heat of Vaporization	32.517 Kg-cal/gm-atom
Valence	+2	Heat of Fusion	2.14 Kg-cal/gm-atom
Subshell Filling	s	Specific Heat	0.25 cal/gm/°C
Atomic Radius	1.60 Å	Thermal Conductance	0.38 cal/cm²/cm/°C/sec
Covalent Radius	1.30 Å	Boiling Point	1107° C
Ionic Radius	0.65 Å(+2)	Melting Point	650° C
Atomic Vol.	14.0 W/D		
Core	Neon		

Ground State Electron Config.
$1s^2\ 2s^2\ 2p^6\ 3s^2$

Radioactive Isotopes
None.

Manganese		Mn	
Principal Quantum No.	4	Valence Electrons	3d⁵ 4s² (+7)
Atomic No.	25	Acid-Base Property	Very acidic
Atomic Wt.	54.9380		
X-Ray Notation	N	Density	7.43 gm/ml
Group	VII B	Crystal Structure	Cubic
Category	Heavy transitional metal	Electronegativity	1.5
State	Solid	Electrical Conductance	0.054 micro-ohm
Origin	Natural	First Ionization Energy	171 Kcal/gm-mole
No. of Protons	25		
No. of Electrons	25	Ionization Potential	7.4 ev
Valence	+2, +4, +7 (also, +3, +6)	Heat of Vaporization	53.7 Kg-cal/gm-atom
Subshell Filling	d	Heat of Fusion	3.50 Kg-cal/gm-atom
		Specific Heat	0.115 cal/gm/°C
Atomic Radius	1.26 Å	Thermal Conductance	/
Covalent Radius	/	Boiling Point	2150° C
Ionic Radius	0.80 Å(+2); 0.46 Å(+7)	Melting Point	1245° C
Atomic Vol.	739 W/D		
Core	Argon		

Ground State Electron Config.
1s² 2s² 2p⁶ 3s² 3p⁶ 3d⁵ 4s²

Radioactive Isotopes
None.

Mendelevium — Md

Principal Quantum No.	7	Valence Electrons	6d^1 7s^2 (+3)
Atomic No.	101	Acid-Base Property	/
Atomic Wt.	256 △	Density	/
Group	VI A	Crystal Structure	/
Category	Actinon	Electro-negativity	/
State	Solid		
Origin	Synthetic		
No. of Protons	101	Electrical Conductance	/
No. of Electrons	101	First Ionization Energy	/
Valence	+3	Ionization Potential	/
Subshell Filling	f	Heat of Vaporization	/
Atomic Radius	/	Heat of Fusion	/
Covalent Radius	/	Specific Heat	/
Ionic Radius	/	Thermal Conductance	/
Atomic Vol.	/	Boiling Point	/
Core	Krypton	Melting Point	/

Ground State Electron Config.
1s^2 2s^2 2p^6 3s^2 3p^6 3d^{10}
4s^2 4p^6 4d^{10} 4f^{14} 5s^2 5p^6
5d^{10} 5f^{12} 5g^0
6s$^{(2}$ 6p^6 6d^1 7s$^{2)}$

Radioactive Isotopes
Md (at.wt.=256); ½life=67 hours
with decay via β^- and γ.
Radioactivity is induced.

	Mercury		Hg (Hydrargyrum)

Principal Quantum No.	6	Valence Electrons	$6s^2$ (+2)
Atomic No.	80	Acid-Base Property	Moderately basic
Atomic Wt.	200.59		
X-Ray Notation	P	Density	13.6 gm/ml
Group	II B	Crystal Structure	Rhombohedral
Category	Heavy transitional metal	Electronegativity	1.9
State	Liquid	Electrical Conductance	0.011 micro-ohm
Origin	Natural	First Ionization Energy	241 Kcal/gm-mole
No. of Protons	80		
No. of Electrons	80	Ionization Potential	10.4 ev
Valence	+1, +2	Heat of Vaporization	13.9 Kg-cal/gm-atom
Subshell Filling	s	Heat of Fusion	0.56 Kg-cal/gm-atom
Atomic Radius	1.57 Å	Specific Heat	0.033 cal/gm/°C
Covalent Radius	1.49 Å	Thermal Conductance	0.02 cal/cm²/cm/°C/sec
Ionic Radius	1.10 Å(+2)	Boiling Point	357° C
Atomic Vol.	14.8 W/D	Melting Point	−38.4° C
Core	Krypton		

Ground State Electron Config.
$1s^2\ 2s^2\ 2p^6\ 3s^2\ 3p^6\ 3d^{10}$
$4s^2\ 4p^6\ 4d^{10}\ 4f^{14}\ 5s^2\ 5p^6$
$5d^{10}\ 5f^0\ 5g^0\ 6s^2$

Radioactive Isotopes
Hg (at.wt.=197); ½life=65 hours
with decay via γ, e⁻ and K.
Hg (at.wt.=203); ½life=47 days
with decay via β⁻, γ and e⁻.
Radioactivity is induced.

Molybdenum   Mo

Principal Quantum No.	5	Valence Electrons	$4d^5\ 5s^1$ (+6)
Atomic No.	42	Acid-Base Property	Acidic
Atomic Wt.	95.94	Density	10.2 gm/ml
X-Ray Notation	Q	Crystal Structure	Cubic—body centered
Group	VI B		
Category	Heavy transitional metal	Electronegativity	1.8
State	Solid	Electrical Conductance	0.19 micro-ohm
Origin	Natural	First Ionization Energy	166 Kcal/gm-mole
No. of Protons	42		
No. of Electrons	42	Ionization Potential	7.2 ev
Valence	+3, +6 (also, +2, +4, +5)	Heat of Vaporization	128 Kg-cal/gm-atom
Subshell Filling	d	Heat of Fusion	6.6 Kg-cal/gm-atom
Atomic Radius	1.39 Å	Specific Heat	0.061 cal/gm/°C
Covalent Radius	/	Thermal Conductance	0.35 cal/cm²/cm/°C/sec
Ionic Radius	0.68 Å(+4); 0.62 Å(+6)	Boiling Point	5560° C
Atomic Vol.	9.4 W/D	Melting Point	2610° C
Core	Krypton		

Ground State Electron Config.
$1s^2\ 2s^2\ 2p^6\ 3s^2\ 3p^6\ 4d^{10}$
$4s^2\ 4p^6\ 4d^5\ 4f^0\ 5s^1$

Radioactive Isotopes
Mo (at.wt.=99); ½life=67 hours
with decay via β^- and e^-.
Radioactivity is induced.

Neodymium   Nd

Principal Quantum No.	6	Valence Electrons	$4f^4$ $5d^0$ $6s^2$	
Atomic No.	60	Acid-Base Property	Moderately basic	
Atomic Wt.	144.24	Density	7.00 gm/ml	
X-Ray Notation	P	Crystal Structure	Hexagonal	
Group	VII B			
Category	Lanthanon	Electro-negativity	1.2	
State	Solid	Electrical Conductance	0.013 micro-ohm	
Origin	Natural			
No. of Protons	60	First Ioniza-tion Energy	145 Kcal/gm-mole	
No. of Electrons	60	Ionization Potential	6.3 ev	
Valence	+3, +4	Heat of Vaporization	69 Kg-cal/gm-atom	
Subshell Filling	f	Heat of Fusion	1.70 Kg-cal/gm-atom	
Atomic Radius	1.81 Å	Specific Heat	0.045 cal/gm/°C	
Covalent Radius	1.64 Å	Thermal Conductance	0.013 cal/cm²/cm/°C/sec	
Ionic Radius	1.08 Å	Boiling Point	3027° C	
Atomic Vol.	20.6 W/D	Melting Point	1024° C	
Core	Krypton			

Ground State Electron Config.
1s² 2s² 2p⁶ 3s² 3p⁶ 3d¹⁰
4s² 4p⁶ 4d¹⁰ 4f⁴ 5s² 5p⁶
5d⁰ 5f⁰ 5g⁰ 6s²

Radioactive Isotopes
Nd (at.wt.=147); ½life=11.1 days
with decay via β⁻ and γ.
Radioactivity is induced.

67

Neon   Ne

Principal Quantum No.	2	Valence Electrons	/
Atomic No.	10	Acid-Base Property	/
Atomic Wt.	20.183	Density	1.20 gm/ml
X-Ray Notation	L	Crystal Structure	Cubic—face centered
Group	Inert gas	Electronegativity	/
Category	Inert gas	Electrical Conductance	/
State	Gas		
Origin	Natural		
No. of Protons	10	First Ionization Energy	497 Kcal/gm-mole
No. of Electrons	10	Ionization Potential	21.6 ev
Valence	0	Heat of Vaporization	0.431 Kg-cal/gm-atom
Subshell Filling	p	Heat of Fusion	0.080 Kg-cal/gm-atom
Atomic Radius	/	Specific Heat	/
Covalent Radius	1.31 Å	Thermal Conductance	0.0001 cal/cm^2/cm/°C/sec
Ionic Radius	/	Boiling Point	−246° C
Atomic Vol.	16.8 W/D	Melting Point	−248.6° C
Core	/		

Ground State Electron Config.
1s^2 2s^2 2p^6

Radioactive Isotopes
None.

Neptunium   Np

Principal Quantum No.	7	Valence Electrons	$5f^5$ $6d^0$ $7s^2$
Atomic No.	93	Acid-Base Property	Amphoteric
Atomic Wt.	237 △	Density	19.5 gm/ml
X-Ray Notation	Q	Crystal Structure	/
Group	VIII B		
Category	Actinon	Electro-negativity	1.3
State	Solid	Electrical	
Origin	Synthetic	Conductance	/
No. of Protons	93	First Ionization Energy	/
No. of Electrons	93	Ionization Potential	/
Valence	+2, +3, +4, +5, +6	Heat of Vaporization	/
Subshell Filling	f	Heat of Fusion	/
Atomic Radius	/	Specific Heat	/
Covalent Radius	/	Thermal Conductance	/
Ionic Radius	1.09 Å(+3); 0.88 Å(+4)	Boiling Point	/
		Melting Point	637° C
Atomic Vol.	21.1 W/D		
Core	Krypton		

Ground State Electron Config.
$1s^2$ $2s^2$ $2p^6$ $3s^2$ $3p^6$ $3d^{10}$
$4s^2$ $4p^6$ $4d^{10}$ $4f^{14}$ $5s^2$ $5p^6$
$5d^{10}$ $5f^5$ $5g^0$ $6s^2$ $6p^6$ $6d^0$
$7s^2$

Radioactive Isotopes
Np (at.wt.=237); ½life=2.2×10⁶ years
with decay via α and γ.
Np (at.wt.=239); ½life=2.33 days
with decay via β^- and γ.
Radioactivity is induced.

Nickel   Ni

Principal Quantum No.	4	Valence Electrons	$3d^8$ $4s^2$
Atomic No.	28	Acid-Base Property	Moderately basic
Atomic Wt.	58.71	Density	8.9 gm/ml
X-Ray Notation	N	Crystal Structure	Cubic—face centered
Group	VIII	Electro-negativity	
Category	Heavy transitional metal	Electrical Conductance	0.145 micro-ohm
State	Solid	First Ionization Energy	176 Kcal/gm-mole
Origin	Natural		
No. of Protons	28	Ionization Potential	7.6 ev
No. of Electrons	28	Heat of Vaporization	91.0 Kg-cal/gm-atom
Valence	+2 (also, +1, +3, +4)	Heat of Fusion	4.21 Kg-cal/gm-atom
Subshell Filling	d	Specific Heat	0.105
Atomic Radius	1.24 Å	Thermal Conductance	0.22 cal/cm²/cm/°C/sec
Covalent Radius	/	Boiling Point	2730° C
Ionic Radius	0.78 Å(+2); 0.62 Å(+3)	Melting Point	1453° C
Atomic Vol.	6.6 W/D		
Core	Argon		

Ground State Electron Config.
$1s^2$ $2s^2$ $2p^6$ $3s^2$ $3p^6$ $3d^8$ $4s^2$

Radioactive Isotopes
Ni (at.wt.=63); ½life=125 years with decay via β^-.
Ni (at.wt.=59); ½life=8×10^4 years with decay via K.
Radioactivity is induced.

Niobium   Nb

Principal Quantum No.	5	Valence Electrons	$4d^4\ 5s^1$ (+5)
Atomic No.	41	Acid-Base Property	Moderately acidic
Atomic Wt.	92.906	Density	8.4 gm/ml
X-Ray Notation	Q	Crystal Structure	Cubic—body centered
Group	V B		
Category	Heavy transitional metal	Electronegativity	1.6
State	Solid	Electrical Conductance	0.080 micro-ohm
Origin	Natural	First Ionization Energy	156 Kcal/gm-mole
No. of Protons	41		
No. of Electrons	41	Ionization Potential	6.8 ev
Valence	+3, +5 (also, +2, +4)	Heat of Vaporization	1.6 Kg-cal/gm-atom
Subshell Filling	d	Heat of Fusion	6.4 Kg-cal/gm-atom
Atomic Radius	1.46 Å	Specific Heat	0.065 cal/gm/°C
Covalent Radius	/	Thermal Conductance	0.125 cal/cm²/cm/°C/sec
Ionic Radius	0.70 Å	Boiling Point	3300° C
Atomic Vol.	10.8 W/D	Melting Point	2415° C
Core	Krypton		

Ground State Electron Config.
1s^2 2s^2 2p^6 3s^2 3p^6 3d^{10}
4s^2 4p^6 4d^4 4f^0 5s^1

Radioactive Isotopes
None.

Nitrogen   N

Principal Quantum No.	2		Valence Electrons	$2p^3$
Atomic No.	7		Acid-Base Property	Very acidic
Atomic Wt.	14.0067		Density	0.81 gm/ml
X-Ray Notation	L		Crystal Structure	Hexagonal
Group	V A		Electro-negativity	3.0
Category	Non-metal			
State	Gas		Electrical Conductance	/
Origin	Natural			
No. of Protons	7		First Ionization Energy	336 Kcal/gm-mole
No. of Electrons	7		Ionization Potential	14.5 ev
Valence	−3, −2, −2, +2, +3, +4, +5 (also, +1)		Heat of Vaporization	0.666 Kg-cal/gm-atom
Subshell Filling	p		Heat of Fusion	0.086 Kg-cal/gm-atom
			Specific Heat	0.247
Atomic Radius	0.92 Å		Thermal Conductance	0.00006 cal/cm^2/cm/°C/sec
Covalent Radius	0.75 Å		Boiling Point	−195.8° C
Ionic Radius	1.71 Å(−3); 0.11 Å(+5)		Melting Point	−210° C
Atomic Vol.	17.3 W/D			
Core	/			

Ground State Electron Config.
$1s^2\ 2s^2\ 2p^3$

Radioactive Isotopes
None.

Nobelium   No

Principal Quantum No.	7	Valence Electrons	$6d^1$ $7s^2$
Atomic No.	102	Acid-Base Property	/
Atomic Wt.	254 △	Density	/
X-Ray Notation	Q	Crystal Structure	/
Group	VII A		
Category	Actinon	Electro-negativity	/
State	Solid	Electrical Conductance	/
Origin	Synthetic		
No. of Protons	102	First Ionization Energy	/
No. of Electrons	102	Ionization Potential	/
Valence	+3	Heat of Vaporization	/
Subshell Filling	f	Heat of Fusion	/
Atomic Radius	/	Specific Heat	/
Covalent Radius	/	Thermal Conductance	/
Ionic Radius	/	Boiling Point	/
Atomic Vol.	/	Melting Point	/
Core	Krypton		

Ground State Electron Config.
$1s^2$ $2s^2$ $2p^6$ $3s^2$ $3p^6$ $3d^{10}$ $4s^2$
$4p^6$ $4d^{10}$ $4f^{14}$ $5s^2$ $5p^6$ $5d^{10}$
$5f^{13}$ $5g^0$ $6s^{(2}$ $6p^6$ $6d^1$ $7s^{2)}$

Radioactive Isotopes
None.

Osmium   Os

Principal Quantum No.	6	Valence Electrons	$5d^6$ $6s^2$
Atomic No.	76	Acid-Base Property	Moderately basic
Atomic Wt.	190.2		
X-Ray Notation	P	Density	22.6 gm/ml
Group	VIII	Crystal Structure	Hexagonal
Category	Heavy transitional metal	Electronegativity	2.2
State	Solid	Electrical Conductance	0.11 micro-ohm
Origin	Natural	First Ionization Energy	201 Kcal/gm-mole
No. of Protons	76	Ionization Potential	8.7 ev
No. of Electrons	76	Heat of Vaporization	162 Kg-cal/gm-atom
Valence	+3, +4, +6, +8 (also, +2)	Heat of Fusion	6.4 Kg-cal/gm-atom
Subshell Filling	d	Specific Heat	0.031 cal/gm/°C
Atomic Radius	1.35 Å	Thermal Conductance	/
Covalent Radius	/	Boiling Point	5500° C
Ionic Radius	0.76 Å	Melting Point	2700° C
Atomic Vol.	8.43 W/D		
Core	Krypton		

Ground State Electron Config.
$1s^2$ $2s^2$ $2p^6$ $3s^2$ $3p^6$ $3d^{10}$
$4s^2$ $4p^6$ $4d^{10}$ $4f^{14}$ $5s^2$ $5p^6$
$5d^6$ $5f^0$ $5g^0$ $6s^2$

Radioactive Isotopes
Os (at.wt.=191); ½life=15 days
with decay via β^-, γ and e^-.
Radioactivity is induced.

Oxygen   O

Principal Quantum No.	2		Valence Electrons	$2p^4$ (-2)
Atomic No.	8		Acid-Base Property	/
Atomic Wt.	15.9994		Density	1.14 gm/ml
X-Ray Notation	L		Crystal Structure	Cubic
Group	VI A			
Category	Gas		Electro-negativity	3.5
State	Gas		Electrical Conductance	/
Origin	Natural			
No. of Protons	8		First Ioniza-tion Energy	314 Kcal/gm-mole
No. of Electrons	8		Ionization Potential	13.6 ev
Valence	−2		Heat of Vaporization	0.815 Kg-cal/gm-atom
Subshell Filling	p		Heat of Fusion	0.053 Kg-cal/gm-atom
Atomic Radius	/		Specific Heat	0.218 cal/gm/°C
Covalent Radius	0.73 Å		Thermal Conductance	0.00006 cal/cm²/cm/°C/sec
Ionic Radius	1.40 Å(−2); 0.09 Å(+6)		Boiling Point	−183° C
			Melting Point	−218.8° C
Atomic Vol.	14.0 W/D			
Core	/			

Ground State Electron Config.
1s² 2s² 2p⁴

Radioactive Isotopes
None.

Palladium   Pd

Principal Quantum No.	5		Valence Electrons	4d^{10}
Atomic No.	46		Acid-Base Property	Moderately basic
Atomic Wt.	106.14		Density	12.0 gm/ml
X-Ray Notation	Q		Crystal Structure	Cubic—face centered
Group	VIII		Electronegativity	2.2
Category	Heavy transitional metal		Electrical Conductance	0.093 micro-ohm
State	Solid		First Ionization Energy	192 Kcal/gm-mole
Origin	Natural		Ionization Potential	8.3 ev
No. of Protons	46		Heat of Vaporization	90 Kg-cal/gm-atom
No. of Electrons	46		Heat of Fusion	4.0 Kg-cal/gm-atom
Valence	+2, +4		Specific Heat	0.058 cal/gm/°C
Subshell Filling	d		Thermal Conductance	0.17 cal/cm^2/cm/°C/sec
Atomic Radius	1.37 Å		Boiling Point	3980° C
Covalent Radius	/		Melting Point	1552° C
Ionic Radius	0.50 Å			
Atomic Vol.	8.9 W/D			
Core	Krypton			

Ground State Electron Config.
1s^2 2s^2 2p^6 3s^2 3p^6 3d^{10}
4s^2 4p^6 4d^{10} 4f^0

Radioactive Isotopes
Pd (at.wt.=103); ½life=17 days
with decay via γ and K.
Radioactivity is induced.

Phosphorus   P

Principal Quantum No.	3	Valence Electrons	$3p^3$
Atomic No.	15	Acid-Base Property	Moderately acidic
Atomic Wt.	30.9738	Density	1.82 gm/ml
X-Ray Notation	M	Crystal Structure	Cubic
Group	V A		
Category	Non-metal	Electro-negativity	2.1
State	Solid	Electrical Conductance	10^{-7} micro-ohm
Origin	Natural		
No. of Protons	15	First Ioniza-tion Energy	254 Kcal/gm-mole
No. of Electrons	15	Ionization Potential	11.0 ev
Valence	+1, +3, +5, −3	Heat of Vaporization	2.97 Kg-cal/gm-atom
Subshell Filling	p	Heat of Fusion	0.15 Kg-cal/gm-atom
Atomic Radius	1.28 Å	Specific Heat	0.177 cal/gm/°C
Covalent Radius	1.06 Å	Thermal Conductance	/
Ionic Radius	2.12 Å(−3); 0.34 Å(+5)	Boiling Point	280° C
		Melting Point	44.2° C
Atomic Vol.	17.0 W/D		
Core	Neon		

Ground State Electron Config.
$1s^2\ 2s^2\ 2p^6\ 3s^2\ 3p^3$

Radioactive Isotopes
P (at.wt.=32); ½life=14.2 days
with decay via β^-.
Radioactivity is induced.

Platinum		Pt	
Principal Quantum No.	6	Valence Electrons	$5d^9\ 6s^1$
Atomic No.	78	Acid-Base Property	Moderately basic
Atomic Wt.	195.09	Density	21.4 gm/ml
X-Ray Notation	P	Crystal Structure	Cubic—body centered
Group	VIII	Electro-negativity	2.2
Category	Heavy transitional metal	Electrical Conductance	0.095 micro-ohm
State	Solid	First Ionization Energy	207 Kcal/gm-mole
Origin	Natural	Ionization Potential	9.0 ev
No. of Protons	78	Heat of Vaporization	122 Kg-cal/gm-atom
No. of Electrons	78	Heat of Fusion	5.2 Kg-cal/gm-atom
Valence	+2, +4, +6 (also, +1, +3)	Specific Heat	0.032 cal/gm/°C
Subshell Filling	d	Thermal Conductance	0.17 $cal/cm^2/cm/°C/sec$
Atomic Radius	1.38 Å	Boiling Point	4530° C
Covalent Radius	/	Melting Point	1769° C
Ionic Radius	0.52 Å(+2)		
Atomic Vol.	9.10 W/D		
Core	Krypton		

Ground State Electron Config.
$1s^2\ 2s^2\ 2p^6\ 3s^2\ 3p^6\ 3d^{10}$
$4s^2\ 4p^6\ 4d^{10}\ 4f^{14}\ 5s^2\ 5p^6$
$5d^9\ 5f^0\ 5g^0\ 6s^1$

Radioactive Isotopes
Pt (at.wt.=197); ½life=18 hours with decay via β^- and γ.
Radioactivity is induced.

Plutonium   Pu

Principal Quantum No.	7	Valence Electrons	$5f^6\ 7s^2$
Atomic No.	94	Acid-Base Property	Amphoteric
Atomic Wt.	244 △	Density	19.5 gm/ml
X-Ray Notation	Q	Crystal Structure	/
Group	VIII B	Electronegativity	/
Category	Actinon	Electrical Conductance	/
State	Solid		
Origin	Synthetic		
No. of Protons	94	First Ionization Energy	/
No. of Electrons	94	Ionization Potential	/
Valence	+2, +3, +4, +5, +6	Heat of Vaporization	/
Subshell Filling	f	Heat of Fusion	/
Atomic Radius	/	Specific Heat	/
Covalent Radius	/	Thermal Conductance	/
		Boiling Point	/
Ionic Radius	1.07 Å(+3); 0.86 Å(+4)	Melting Point	637° C
Atomic Vol.	/		
Core	Krypton		

Ground State Electron Config.
$1s^2\ 2s^2\ 2p^6\ 3s^2\ 3p^6\ 3d^{10}$
$4s^2\ 4p^6\ 4d^{10}\ 4f^{14}\ 5s^2\ 5p^6$
$5d^{10}\ 5f^6\ 5g^0\ 6s^2\ 6p^6\ 6d^0$
$7s^2$

Radioactive Isotopes
Pu (at.wt.=239); ½life=24,300 years with decay via α and γ; spontaneous fission. This isotope occurs naturally.
Pu (at.wt.=242); ½life=3.8×10^5 years with decay via α; spontaneous fission. Radioactivity is induced.
Pu (at.wt.=241); ½life=13 years with decay via α, γ and β^-. Radioactivity is induced.

Polonium   Po

Principal Quantum No.	6		Valence Electrons	6p⁴
Atomic No.	84		Acid-Base Property	Amphoteric
Atomic Wt.	210 △		Density	9.2 gm/ml
X-Ray Notation	P		Crystal Structure	Monoclinic
Group	VI A		Electro-negativity	2.0
Category	Semi-metal		Electrical Conductance	0.02 micro-ohm
State	Solid			
Origin	Natural		First Ioniza-tion Energy	/
No. of Protons	84		Ionization Potential	/
No. of Electrons	84		Heat of Vaporization	29 Kg-cal/gm-atom
Valence	+2, +4 (also, −2, +6)		Heat of Fusion	/
Subshell Filling	p		Specific Heat	/
Atomic Radius	1.76 Å		Thermal Conductance	/
Covalent Radius	/		Boiling Point	/
Ionic Radius	/		Melting Point	254° C
Atomic Vol.	22.7 W/D			
Core	Krypton			

Ground State Electron Config.
1s² 2s² 2p⁶ 3s² 3p⁶ 3d¹⁰
4s² 4p⁶ 4d¹⁰ 4f¹⁴ 5s² 5p⁶
5d¹⁰ 5f⁰ 5g⁰ 6s² 6p⁴

Radioactive Isotopes
Po (at.wt.=210); ½life=138.4 days with decay via α and γ. This isotope occurs naturally.
Po (at.wt.=209); ½life=100 years with decay via α, γ and K. Radioactivity is induced.

Potassium   K

Principal Quantum No.	4	Valence Electrons	$4s^1$
Atomic No.	19	Acid-Base Property	Very basic
Atomic Wt.	39.102	Density	0.86 gm/ml
X-Ray Notation	N	Crystal Structure	Cubic—body centered
Group	I A	Electro-negativity	0.8
Category	Alkali earth metal	Electrical Conductance	0.143 micro-ohm
State	Solid	First Ioniza-tion Energy	100 Kcal/gm-mole
Origin	Natural		
No. of Protons	19	Ionization Potential	4.3 ev
No. of Electrons	19	Heat of Vaporization	18.9 Kg-cal/gm-atom
Valence	+1		
Subshell Filling	s	Heat of Fusion	0.55 Kg-cal/gm-atom
Atomic Radius	2.35 Å	Specific Heat	0.177 cal/gm/°C
Covalent Radius	1.96 Å	Thermal Conductance	0.23 cal/cm²/cm/°C/sec
Ionic Radius	1.33 Å(+1)	Boiling Point	760° C
Atomic Vol.	45.3 W/D	Melting Point	63.7° C
Core	Argon		

Ground State Electron Config.
$1s^2\ 2s^2\ 2p^6\ 3s^2\ 3p^6\ 3d^0\ 4s^1$

Radioactive Isotopes
K (at.wt.=40); ½life=10^9 years with decay via β^+, γ and K. This isotope occurs naturally.
K (at.wt.=42); ½life=12.4 hours with decay via β^- and γ. Radioactivity is induced.

81

Praseodymium   Pr

Principal Quantum No.	6		Valence Electrons	$4f^3\ 6s^2$
Atomic No.	59		Acid-Base Property	Moderately basic
Atomic Wt.	140.907			
X-Ray Notation	P		Density	6.77 gm/ml
Group	VI B		Crystal Structure	Hexagonal
Category	Lanthanon		Electro-negativity	1.1
State	Solid			
Origin	Natural		Electrical Conductance	0.015 micro-ohm
No. of Protons	59		First Ioniza-tion Energy	133 Kcal/gm-mole
No. of Electrons	59		Ionization Potential	5.8 ev
Valence	+3		Heat of Vaporization	79 Kg-cal/gm-atom
Subshell Filling	f		Heat of Fusion	1.60 Kg-cal/gm-atom
Atomic Radius	1.82 Å		Specific Heat	0.048 cal/gm/°C
Covalent Radius	1.65 Å		Thermal Conductance	0.028 cal/cm²/cm/°C/sec
Ionic Radius	1.09 Å(+3); 0.92 Å(+4)		Boiling Point	3127° C
Atomic Vol.	20.8 W/D		Melting Point	935° C
Core	Krypton			

Ground State Electron Config.
$1s^2\ 2s^2\ 2p^6\ 3s^2\ 3p^6\ 3d^{10}$
$4s^2\ 4p^6\ 4d^{10}\ 4f^3\ 5s^2\ 5p^6$
$5d^0\ 5f^0\ 5g^0\ 6s^2$

Radioactive Isotopes
Pr (at.wt.=143); ½life=13.8 days
with decay via β^-.
Radioactivity is induced.

Promethium   Pm

Principal Quantum No.	6	Valence Electrons	$4f^5\ 6s^2$
Atomic No.	61	Acid-Base Property	Moderately basic
Atomic Wt.	145 △	Density	/
X-Ray Notation	P	Crystal Structure	Hexagonal
Group	VIII B	Electro-negativity	/
Category	Lanthanon		
State	Solid	Electrical Conductance	/
Origin	Synthetic		
No. of Protons	61	First Ionization Energy	/
No. of Electrons	61	Ionization Potential	/
Valence	+3	Heat of Vaporization	/
Subshell Filling	f	Heat of Fusion	/
Atomic Radius	/	Specific Heat	/
Covalent Radius	/	Thermal Conductance	/
Ionic Radius	1.06 Å(+3)	Boiling Point	/
Atomic Vol.	/	Melting Point	1027° C
Core	Krypton		

Ground State Electron Config.
$1s^2\ 2s^2\ 2p^6\ 3s^2\ 3p^6\ 3d^{10}$
$4s^2\ 4p^6\ 4d^{10}\ 4f^5\ 5s^2\ 5p^6$
$5d^0\ 5f^0\ 5g^0\ 6s^2$

Radioactive Isotopes
Pm (at.wt.=147); ½life=2.6 years
with decay via β^-.
Pm (at.wt.=145); ½life=25 years
with decay via α, K and L.
Radioactivity is induced.

Protractinium   Pa

Principal Quantum No.	7	Valence Electrons	$5f^2\ 6d^1\ 7s^2$
Atomic No.	91	Acid-Base Property	Slightly basic
Atomic Wt.	231 △	Density	15.4 gm/ml
X-Ray Notation	Q	Crystal Structure	/
Group	VI B		
Category	Actinon	Electro-negativity	1.5
State	Solid	Electrical Conductance	/
Origin	Natural		
No. of Protons	91	First Ionization Energy	/
No. of Electrons	91	Ionization Potential	/
Valence	+5, +4, +3	Heat of Vaporization	130Kg-cal/gm-atom
Subshell Filling	f	Heat of Fusion	/
Atomic Radius	/	Specific Heat	/
Covalent Radius	/	Thermal Conductance	/
Ionic Radius	1.12 Å(+3); 0.91 Å(+4)	Boiling Point	/
Atomic Vol.	15.0 W/D	Melting Point	1230° C
Core	Krypton		

Ground State Electron Config.
$1s^2\ 2s^2\ 2p^6\ 3s^2\ 3p^6\ 3d^{10}$
$4s^2\ 4p^6\ 4d^{10}\ 4f^{14}\ 5s^2\ 5p^6$
$5d^{10}\ 5f^2\ 5g^0\ 6s^2\ 6p^6\ 6d^1$
$7s^2$

Radioactive Isotopes
Pa (at.wt.=231); ½life=34,000 years
with decay via α and γ.
This isotope occurs naturally.

Radium   Ra

Principal Quantum No.	7	Valence Electrons	$7s^2$ (+2)
Atomic No.	88	Acid-Base Property	Very basic
Atomic Wt.	226 △	Density	5.0 gm/ml
X-Ray Notation	Q	Crystal Structure	/
Group	II A		
Category	Alkali earth metal	Electro-negativity	0.9
State	Solid	Electrical Conductance	/
Origin	Natural	First Ioniza-tion Energy	/
No. of Protons	88		
No. of Electrons	88	Ionization Potential	5.3 ev
Valence	+2	Heat of Vaporization	27.4 Kg-cal/gm-atom
Subshell Filling	s	Heat of Fusion	2.4 Kg-cal/gm-atom
Atomic Radius	/	Specific Heat	/
Covalent Radius	/	Thermal Conductance	/
Ionic Radius	1.40 Å(+2)	Boiling Point	/
Atomic Vol.	45 W/D	Melting Point	700° C
Core	Krypton		

Ground State Electron Config.
$1s^2$ $2s^2$ $2p^6$ $3s^2$ $3p^6$ $3d^{10}$
$4s^2$ $4p^6$ $4d^{10}$ $4f^{14}$ $5s^2$ $5p^6$
$5d^{10}$ $5f^0$ $5g^0$ $6s^2$ $6p^6$ $7s^2$

Radioactive Isotopes
Ra (at.wt.=226); ½life=1620 years
with decay via α and γ.
This isotope occurs naturally.

Radon   Rn

Principal Quantum No.	6	Valence Electrons	$6p^6$	
Atomic No.	86	Acid-Base Property	/	
Atomic Wt.	222 △	Density	/	
X-Ray Notation	P	Crystal Structure	Cubic—face centered	
Group	Inert gas			
Category	Inert gas	Electro-negativity	/	
State	Gas	Electrical Conductance	/	
Origin	Natural			
No. of Protons	86	First Ionization Energy	248 Kcal/gm-mole	
No. of Electrons	86	Ionization Potential	10.7 ev	
Valence	0	Heat of Vaporization	3.92 Kg-cal/gm-atom	
Subshell Filling	p	Heat of Fusion	0.69 Kg-cal/gm-atom	
Atomic Radius	/	Specific Heat	/	
Covalent Radius	2.14 Å	Thermal Conductance	/	
Ionic Radius	/	Boiling Point	−61.8° C	
Atomic Vol.	/	Melting Point	−71° C	
Core	Krypton			

Ground State Electron Config.
$1s^2$ $2s^2$ $2p^6$ $3s^2$ $3p^6$ $3d^{10}$
$4s^2$ $4p^6$ $4d^{10}$ $4f^{14}$ $5s^2$ $5p^6$
$5d^{10}$ $5f^0$ $5g^0$ $6s^2$ $6p^6$

Radioactive Isotopes
Rn (at.wt.=222); ½life=3.82 days
with decay via α.
This isotope occurs naturally.

Rhenium   Re

Principal Quantum No.	6	
Atomic No.	75	
Atomic Wt.	186.2	
X-Ray Notation	P	
Group	VII B	
Category	Heavy transitional metal	
State	Solid	
Origin	Natural	
No. of Protons	75	
No. of Electrons	75	
Valence	+2, +4, +6, +7 (also, +1, +3, +5, −1)	
Subshell Filling	d	
Atomic Radius	1.37 Å	
Covalent Radius	/	
Ionic Radius	/	
Atomic Vol.	8.85 W/D	
Core	Krypton	

Valence Electrons	$5d^5\ 6s^2$
Acid-Base Property	Moderately acidic
Density	21.0 gm/ml
Crystal Structure	Hexagonal
Electronegativity	1.9
Electrical Conductance	0.051 micro-ohm
First Ionization Energy	182 Kcal/gm-mole
Ionization Potential	7.9 ev
Heat of Vaporization	152 Kg-cal/gm-atom
Heat of Fusion	7.9 Kg-cal/gm-atom
Specific Heat	0.033 cal/gm/°C
Thermal Conductance	0.17 cal/cm²/cm/°C/sec
Boiling Point	5900° C
Melting Point	3180° C

Ground State Electron Config.
$1s^2\ 2s^2\ 2p^6\ 3s^2\ 3p^6\ 3d^{10}$
$4s^2\ 4p^6\ 4d^{10}\ 4f^{14}\ 5s^2\ 5p^6$
$5d^5\ 5f^0\ 5g^0\ 6s^2$

Radioactive Isotopes
Re (at.wt.=188); ½life=16.7 hours
with decay via β^- and γ.
Re (at.wt.=186); ½life=3.7 days
with decay via β^- and γ.
Radioactivity is induced.

Rhodium   Rh

Principal Quantum No.	5	Valence Electrons	$4d^8$ $5s^1$
Atomic No.	45	Acid-Base Property	Amphoteric
Atomic Wt.	102.905	Density	12.4 gm/ml
X-Ray Notation	Q	Crystal Structure	Cubic—face centered
Group	VIII	Electronegativity	2.2
Category	Heavy transitional metal	Electrical Conductance	0.22 micro-ohm
State	Solid	First Ionization Energy	178 Kcal/gm-mole
Origin	Natural	Ionization Potential	7.7 ev
No. of Protons	45	Heat of Vaporization	127 Kg-cal/gm-atom
No. of Electrons	45	Heat of Fusion	5.2 Kg-cal/gm-atom
Valence	+4 (also, +2, +3, +6)	Specific Heat	0.059 cal/gm/°C
Subshell Filling	d	Thermal Conductance	0.21 cal/cm²/cm/°C/sec
Atomic Radius	1.43 Å	Boiling Point	4500° C
Covalent Radius	/	Melting Point	1966° C
Ionic Radius	0.86 Å (+2)		
Atomic Vol.	8.3 W/D		
Core	Krypton		

Ground State Electron Config.
$1s^2$ $2s^2$ $2p^6$ $3s^2$ $3p^6$ $3d^{10}$
$4s^2$ $4p^6$ $4d^8$ $4f^0$ $5s^1$

Radioactive Isotopes
None.

Rubidium   Rb

Principal Quantum No.	5	Valence Electrons	$5s^1$
Atomic No.	37	Acid-Base Property	Very basic
Atomic Wt.	85.47	Density	1.53 gm/ml
X-Ray Notation	Q	Crystal Structure	Cubic—body centered
Group	I A	Electro-negativity	0.8
Category	Alkali earth metal	Electrical Conductance	0.080 micro-ohm
State	Solid	First Ioniza-tion Energy	96 Kcal/gm-mole
Origin	Natural		
No. of Protons	37	Ionization Potential	4.2 ev
No. of Electrons	37	Heat of Vaporization	18.1 Kg-cal/gm-atom
Valence	+1		
Subshell Filling	s	Heat of Fusion	0.55 Kg-cal/gm-atom
Atomic Radius	2.48 Å	Specific Heat	0.080 cal/gm/°C
Covalent Radius	2.11 Å	Thermal Conductance	/
Ionic Radius	1.48 Å	Boiling Point	688° C
Atomic Vol.	55.9 W/D	Melting Point	38.9° C
Core	Krypton		

Ground State Electron Config.
$1s^2\ 2s^2\ 2p^6\ 3s^2\ 3p^6\ 3d^{10}$
$4s^2\ 4p^6\ 4d^0\ 4f^0\ 5s^1$

Radioactive Isotopes
Rb (at.wt.=86); ½life=18.6 hours
with decay via β^- and γ.
Radioactivity is induced.

Ruthenium **Ru**

Principal Quantum No.	5
Atomic No.	44
Atomic Wt.	101.07
X-Ray Notation	Q
Group	VIII
Category	Heavy transitional metal
State	Solid
Origin	Natural
No. of Protons	44
No. of Electrons	44
Valence	+3, +4, +8 (also, +2, +6, +7)
Subshell Filling	d
Atomic Radius	1.34 Å
Covalent Radius	/
Ionic Radius	0.69 Å(+3); 0.65 Å(+4)
Atomic Vol.	8.3 W/D
Core	Krypton

Valence Electrons	$4d^7\ 5s^1$
Acid-Base Property	Moderately acidic
Density	12.2 gm/ml
Crystal Structure	Hexagonal
Electronegativity	2.2
Electrical Conductance	0.10 micro-ohm
First Ionization Energy	173 Kcal/gm-mole
Ionization Potential	7.5 ev
Heat of Vaporization	148 Kg-cal/gm-atom
Heat of Fusion	6.1 Kg-cal/gm-atom
Specific Heat	0.057 cal/gm/°C
Thermal Conductance	/
Boiling Point	4900° C
Melting Point	2500° C

Ground State Electron Config.
$1s^2\ 2s^2\ 2p^6\ 3s^2\ 3p^6\ 3d^{10}$
$4s^2\ 4p^6\ 4d^7\ 4f^0\ 5s^1$

Radioactive Isotopes
Ru (at.wt.=103); ½life=40 days with decay via β^- and γ.
Ru (at.wt.=97); ½life=2.9 days with decay via γ, e^- and K.
Radioactivity is induced.

Samarium   Sm

Principal Quantum No.	6	Valence Electrons	$4f^6\ 6s^2$
Atomic No.	62	Acid-Base Property	Slightly basic
Atomic Wt.	150.35	Density	7.54 gm/ml
X-Ray Notation	P	Crystal Structure	Rhombohedral
Group	VIII B		
Category	Lanthanum	Electro-negativity	1.2
State	Solid	Electrical Conductance	0.011 micro-ohm
Origin	Natural		
No. of Protons	62	First Ioniza-tion Energy	129 Kcal/gm-mole
No. of Electrons	62	Ionization Potential	5.6 ev
Valence	+3 (also, +2)	Heat of Vaporization	46 Kg-cal/gm-atom
Subshell Filling	f	Heat of Fusion	2.1 Kg-cal/gm-atom
Atomic Radius	1.66 Å	Specific Heat	0.042 cal/gm/°C
Covalent Radius	1.66 Å	Thermal Conductance	/
Ionic Radius	1.04 Å(+3)	Boiling Point	1900° C
Atomic Vol.	19.9 W/D	Melting Point	1072° C
Core	Krypton		

Ground State Electron Config.
$1s^2\ 2s^2\ 2p^6\ 3s^2\ 3p^6\ 3d^{10}$
$4s^2\ 4p^6\ 4d^{10}\ 4f^6\ 5s^2\ 5p^6$
$5d^0\ 5f^0\ 5g^0\ 6s^2$

Radioactive Isotopes
Sm (at.wt.=153); ½life=47 hours
with decay via β^- and γ.
Sm (at.wt.=145); ½life=340 days
with decay via γ and K.
Radioactivity is induced.

Scandium   Sc

Principal Quantum No.	4		Valence Electrons	$3d^1\ 4s^2$
Atomic No.	21		Acid-Base Property	Moderately basic
Atomic Wt.	44.956		Density	3.0 gm/ml
X-Ray Notation	N		Crystal Structure	Hexagonal
Group	III B			
Category	Heavy transitional metal		Electronegativity	1.3
State	Solid		Electrical Conductance	0.015 micro-ohm
Origin	Natural		First Ionization Energy	151 Kcal/gm-mole
No. of Protons	21		Ionization Potential	6.6 ev
No. of Electrons	21		Heat of Vaporization	81 Kg-cal/gm-atom
Valence	+3		Heat of Fusion	3.8 Kg-cal/gm-atom
Subshell Filling	d		Specific Heat	0.13 cal/gm/°C
Atomic Radius	1.62 Å		Thermal Conductance	0.015 cal/cm²/cm/°C/sec
Covalent Radius	1.44 Å		Boiling Point	2730° C
Ionic Radius	0.81 Å (+3)		Melting Point	1539° C
Atomic Vol.	15.0 W/D			
Core	Argon			

Ground State Electron Config.
$1s^2\ 2s^2\ 2p^6\ 3s^2\ 3p^6\ 3d^1\ 4s^2$

Radioactive Isotopes
Sc (at.wt.=46); ½Life=84 days
with decay via β^- and γ.
Radioactivity is induced.

Selenium   Se

Principal Quantum No.	4	Valence Electrons	$4s^2\ 4p^4$
Atomic No.	34	Acid-Base Property	Very acidic
Atomic Wt.	78.96	Density	4.79 gm/ml
X-Ray Notation	N	Crystal Structure	Hexagonal
Group	VI A	Electro-negativity	2.4
Category	Non-metals		
State	Solid	Electrical Conductance	0.08 micro-ohm
Origin	Natural		
No. of Protons	34	First Ioniza-tion Energy	225 Kcal/gm-mole
No. of Electrons	34	Ionization Potential	9.8 ev
Valence	+4, +6, −3 (also, +2)	Heat of Vaporization	3.34 Kg-cal/gm-atom
Subshell Filling	p	Heat of Fusion	1.25 Kg-cal/gm-atom
Atomic Radius	1.40 Å	Specific Heat	0.084 cal/gm/°C
Covalent Radius	1.16 Å	Thermal Conductance	0.00005 cal/cm²/cm/°C/sec
Ionic Radius	1.98 Å(−2); 0.42 Å(+6)	Boiling Point	685° C
		Melting Point	217° C
Atomic Vol.	16.5 W/D		
Core	Argon		

Ground State Electron Config.
1s² 2s² 2p⁶ 3s² 3p⁶ 3d¹⁰
4s² 4p⁴

Radioactive Isotopes
Se (at.wt.=75); ½life=121 days
with decay via γ and K.
Radioactivity is induced.

Silicon   Si

Principal Quantum No.	3		Valence Electrons	3p²
Atomic No.	14		Acid-Base Property	Amphoteric
Atomic Wt.	28.086			
X-Ray Notation	M		Density	2.33 gm/ml
Group	IV A		Crystal Structure	Diamond
Category	Non-metal		Electronegativity	1.8
State	Solid		Electrical Conductance	0.10 micro-ohm
Origin	Natural			
No. of Protons	14		First Ionization Energy	188 Kcal/gm-mole
No. of Electrons	14		Ionization Potential	8.1 ev
Valence	+4, −4 (also, +2)		Heat of Vaporization	40.0 Kg-cal/gm-atom
Subshell Filling	p		Heat of Fusion	11.1 Kg-cal/gm-atom
Atomic Radius	1.32 Å		Specific Heat	0.162 cal/gm/°C
Covalent Radius	1.11 Å		Thermal Conductance	0.20 cal/cm²/cm/°C/sec
Ionic Radius	2.71 Å(−4); 0.41 Å(+4)		Boiling Point	2680° C
Atomic Vol.	12.1 W/D		Melting Point	1410° C
Core	Neon			

Ground State Electron Config.
1s² 2s² 2p⁶ 3s² 3p²

Radioactive Isotopes
None.

Silver   Ag (Argentum)

Principal Quantum No.	5	Valence Electrons	$5s^1$
Atomic No.	47	Acid-Base Property	Amphoteric
Atomic Wt.	107.8682	Density	10.5 gm/ml
X-Ray Notation	Q	Crystal Structure	Cubic—face centered
Group	I B	Electro-negativity	1.9
Category	Heavy transitional metal	Electrical Conductance	0.616 micro-ohm
State	Solid	First Ionization Energy	175 Kcal/gm-mole
Origin	Natural		
No. of Protons	47	Ionization Potential	7.6 ev
No. of Electrons	47	Heat of Vaporization	60.7 Kg-cal/gm-atom
Valence	+1 (also, +2, +3)	Heat of Fusion	2.70 Kg-cal/gm-atom
Subshell Filling	s	Specific Heat	0.056 cal/gm/°C
Atomic Radius	1.44 Å	Thermal Conductance	0.98 cal/cm²/cm/°C/sec
Covalent Radius	1.53 Å	Boiling Point	2210° C
Ionic Radius	1.26 Å(+1)	Melting Point	960.8° C
Atomic Vol.	10.3 W/D		
Core	Krypton		

Ground State Electron Config.
1s² 2s² 2p⁶ 3s² 3p⁶ 3d¹⁰
4s² 4p⁶ 4d¹⁰ 4f⁰ 5s¹

Radioactive Isotopes
Ag (at.wt.=110); ½life=24 seconds
with decay via β⁻ and γ.
Ag (at.wt.=111); ½life=7.5 days
with decay via β⁻ and γ.
Radioactivity is induced.

Sodium   Na (Natrium)

Principal Quantum No.	3	Valence Electrons	3s^1
Atomic No.	11	Acid-Base Property	Very basic
Atomic Wt.	22.9898	Density	0.97 gm/ml
X-Ray Notation	M	Crystal Structure	Cubic—body centered
Group	I A		
Category	Alkali earth metal	Electro-negativity	0.9
State	Solid	Electrical Conductance	0.218 micro-ohm
Origin	Natural	First Ioniza-tion Energy	119 Kcal/gm-mole
No. of Protons	11	Ionization Potential	5.1 ev
No. of Electrons	11	Heat of Vaporization	24.12 Kg-cal/gm-atom
Valence	+1		
Subshell Filling	s	Heat of Fusion	0.62 Kg-cal/gm-atom
Atomic Radius	1.90 Å	Specific Heat	0.295 cal/gm/°C
Covalent Radius	1.54 Å	Thermal Conductance	0.32 cal/cm^2/cm/°C/sec
Ionic Radius	0.95 Å (+1)	Boiling Point	892° C
Atomic Vol.	23.7 W/D	Melting Point	97.8° C
Core	Neon		

Ground State Electron Config.
1s^2 2s^2 2p^6 3s^1

Radioactive Isotopes
Na (at.wt.=22); ½life=2.6 years
with decay via β^+, γ and K.
Na (at.wt.=24); ½life=15 hours
with decay via β^- and γ.
Radioactivity is induced.

Strontium   Sr

Principal Quantum No.	5	Valence Electrons	5s²
Atomic No.	38	Acid-Base Property	Very basic
Atomic Wt.	87.62	Density	2.6 gm/ml
X-Ray Notation	Q	Crystal Structure	Cubic—face centered
Group	II A	Electro-negativity	1.0
Category	Alkali earth metal	Electrical Conductance	0.043 micro-ohm
State	Solid	First Ioniza-tion Energy	131 Kcal/gm-mole
Origin	Natural	Ionization Potential	5.7 ev
No. of Protons	38	Heat of Vaporization	33.8 Kg-cal/gm-atom
No. of Electrons	38	Heat of Fusion	2.1 Kg-cal/gm-atom
Valence	+2	Specific Heat	0.176 cal/gm/°C
Subshell Filling	d	Thermal Conductance	/
Atomic Radius	2.15 Å	Boiling Point	1380° C
Covalent Radius	1.92 Å	Melting Point	768° C
Ionic Radius	1.13 Å(+2)		
Atomic Vol.	33.7 W/D		
Core	Krypton		

Ground State Electron Config.
$1s^2\ 2s^2\ 2p^6\ 3s^2\ 3p^6\ 3d^{10}$
$4s^2\ 4p^6\ 4d^0\ 4f^0\ 5s^2$

Radioactive Isotopes
Sr (at.wt.=90); ½life=28 years
with decay via β^-.
Sr (at.wt.=89); ½life=51 days
with decay via β^- and γ.
Sr (at.wt.=85); ½life=64 days
with decay via γ and K.
Radioactivity is induced.

97

Sulfur S

Principal Quantum No.	3	Valence Electrons	$3p^4$
Atomic No.	16	Acid-Base Property	Very acidic
Atomic Wt.	32.064	Density	2.07 gm/ml
X-Ray Notation	M	Crystal Structure	Orthorhombic
Group	VI A	Electro-negativity	2.5
Category	Non-metal	Electrical Conductance	10^{-23} micro-ohm
State	Solid	First Ionization Energy	239 Kcal/gm-mole
Origin	Natural	Ionization Potential	10.4 ev
No. of Protons	16	Heat of Vaporization	3.01 Kg-cal/gm-atom
No. of Electrons	16	Heat of Fusion	0.34 Kg-cal/gm-atom
Valence	+2, +4, +6, −2	Specific Heat	0.175 cal/gm/°C
Subshell Filling	p	Thermal Conductance	0.0007 $cal/cm^2/cm/°C/sec$
Atomic Radius	1.27 Å	Boiling Point	444.6° C
Covalent Radius	1.02 Å	Melting Point	119.0° C
Ionic Radius	1.84 Å(−2); 0.29 Å(+6)		
Atomic Vol.	15.5 W/D		
Core	Neon		

Ground State Electron Config.
$1s^2\ 2s^2\ 2p^6\ 3s^2\ 3p^4$

Radioactive Isotopes
S (at.wt.=35); ½life=87 days with decay via β^-.
Radioactivity is induced.

Tantalum   Ta

Principal Quantum No.	6	Valence Electrons	5d³ 6s²	
Atomic No.	73	Acid-Base Property	Moderately acidic	
Atomic Wt.	180.948			
X-Ray Notation	P	Density	16.6 gm/ml	
Group	V B	Crystal Structure	Cubic—body centered	
Category	Heavy transitional metal	Electronegativity	1.5	
State	Solid	Electrical Conductance	0.081 micro-ohm	
Origin	Natural			
No. of Protons	73	First Ionization Energy	138 Kcal/gm-mole	
No. of Electrons	73	Ionization Potential	6.0 ev	
Valence	+5 (also, +3, +4)	Heat of Vaporization	180 Kg-cal/gm-atom	
Subshell Filling	d	Heat of Fusion	6.8 Kg-cal/gm-atom	
Atomic Radius	1.46 Å	Specific Heat	0.036 cal/gm/°C	
Covalent Radius	/	Thermal Conductance	0.13 cal/cm²/cm/°C/sec	
Ionic Radius	0.73 Å	Boiling Point	5425° C	
Atomic Vol.	10.9 W/D	Melting Point	2996° C	
Core	Krypton			

Ground State Electron Config.
1s² 2s² 2p⁶ 3s² 3p⁶ 3d¹⁰
4s² 4p⁶ 4d¹⁰ 4f¹⁴ 5s² 5p⁶
5d³ 5f⁰ 5g⁰ 6s²

Radioactive Isotopes
Ta (at.wt.=182); ½life=115 days
with decay via β⁻ and γ.
Radioactivity is induced.

99

Technetium   Tc

Principal Quantum No.	5	
Atomic No.	43	
Atomic Wt.	99 △	
X-Ray Notation	Q	
Group	VII B	
Category	Heavy transitional metal	
State	Solid	
Origin	Synthetic	
No. of Protons	43	
No. of Electrons	43	
Valence	+7	
Subshell Filling	d	
Atomic Radius	1.36 Å	
Covalent Radius	/	
Ionic Radius	/	
Atomic Vol.	/	
Core	Krypton	

Valence Electrons	$4d^{(5)}\ 5s^{(2)}$
Acid-Base Property	Moderately acidic
Density	11.5 gm/ml
Crystal Structure	/
Electronegativity	1.9
Electrical Conductance	/
First Ionization Energy	167 Kcal/gm-mole
Ionization Potential	/
Heat of Vaporization	120 Kg-cal/gm-atom
Heat of Fusion	5.5 Kg-cal/gm-atom
Specific Heat	/
Thermal Conductance	/
Boiling Point	/
Melting Point	2200° C

Ground State Electron Config.
$1s^2\ 2s^2\ 2p^6\ 3s^2\ 3p^6\ 3d^{10}$
$4s^2\ 4p^6\ 4d^{(5)}\ 4f^0\ 5s^{(2)}$

Radioactive Isotopes
Tc (at.wt.=99); ½life=2×10^5 years
with decay via β^-.
Tc (at.wt.=97); ½life=10^5 years
with decay via K.
Radioactivity is induced.

Tellurium   Te

Principal Quantum No.	5	Valence Electrons	$5p^4$
Atomic No.	52	Acid-Base Property	Moderately acidic
Atomic Wt.	127.60	Density	6.24 gm/ml
X-Ray Notation	Q	Crystal Structure	Hexagonal
Group	VI A	Electronegativity	2.1
Category	Non-metal (semi-metal)	Electrical Conductance	10^{-6} micro-ohm
State	Solid	First Ionization Energy	208 Kcal/gm-mole
Origin	Natural	Ionization Potential	9.0 ev
No. of Protons	52	Heat of Vaporization	11.9 Kg-cal/gm-atom
No. of Electrons	52	Heat of Fusion	4.28 Kg-cal/gm-atom
Valence	+4, +6, −2 (also, +2)	Specific Heat	0.047 cal/gm/°C
Subshell Filling	p	Thermal Conductance	0.014 cal/cm²/cm/°C/sec
Atomic Radius	1.60 Å	Boiling Point	989.8° C
Covalent Radius	1.35 Å	Melting Point	449.5° C
Ionic Radius	2.21 Å(−2); 0.56 Å(+6)		
Atomic Vol.	20.5 W/D		
Core	Krypton		

Ground State Electron Config.
1s² 2s² 2p⁶ 3s² 3p⁶ 3d¹⁰
4s² 4p⁶ 4d¹⁰ 4f⁰ 5s² 5p⁴

Radioactive Isotopes
Te (at.wt.=127); ½life=9.3 hours
with decay via β⁻.
Radioactivity is induced.

Terbium   Tb

Principal Quantum No.	6	Valence Electrons	$4f^8$ $5d^1$ $6s^2$
Atomic No.	65	Acid-Base Property	Slightly basic
Atomic Wt.	158.924	Density	8.27 gm/ml
X-Ray Notation	P	Crystal Structure	Hexagonal
Group	II B		
Category	Lanthanon	Electro-negativity	1.2
State	Solid	Electrical Conductance	0.009 micro-ohm
Origin	Natural		
No. of Protons	65	First Ionization Energy	155 Kcal/gm-mole
No. of Electrons	65	Ionization Potential	6.7 ev
Valence	+3, +4	Heat of Vaporization	70 Kg-cal/gm-atom
Subshell Filling	f	Heat of Fusion	3.9 Kg-cal/gm-atom
Atomic Radius	1.77 Å	Specific Heat	0.044
Covalent Radius	1.59 Å	Thermal Conductance	/
Ionic Radius	1.00 Å(+3)	Boiling Point	2800° C
Atomic Vol.	19.2 W/D	Melting Point	1356° C
Core	Krypton		

Ground State Electron Config.
$1s^2$ $2s^2$ $2p^6$ $3s^2$ $3p^6$ $3d^{10}$
$4s^2$ $4p^6$ $4d^{10}$ $4f^8$ $5s^2$ $5p^6$
$5d^1$ $5f^0$ $5g^0$ $6s^2$

Radioactive Isotopes
Tb (at.wt.=160); ½life=73 days
with decay via β^- and γ.
Radioactivity is induced.

Thallium   Tl

Principal Quantum No.	6		Valence Electrons	6p^1
Atomic No.	81		Acid-Base Property	Moderately basic
Atomic Wt.	204.37		Density	11.85 gm/ml
X-Ray Notation	P		Crystal Structure	Hexagonal
Group	III A		Electro-negativity	1.8
Category	Heavy transitional metal		Electrical Conductance	0.055 micro-ohm
State	Solid		First Ionization Energy	141 Kcal/gm-mole
Origin	Natural		Ionization Potential	6.1 ev
No. of Protons	81		Heat of Vaporization	38.8 Kg-cal/gm-atom
No. of Electrons	81		Heat of Fusion	1.02 Kg-cal/gm-atom
Valence	+1, +3 (also, +2)		Specific Heat	0.031 cal/gm/°C
Subshell Filling	p		Thermal Conductance	0.093 cal/cm^2/cm/°C/sec
Atomic Radius	1.71 Å		Boiling Point	1457° C
Covalent Radius	1.48 Å		Melting Point	303° C
Ionic Radius	1.40 Å(+1); 0.95 Å(+3)			
Atomic Vol.	17.2 W/D			
Core	Krypton			

Ground State Electron Config.
1s^2 2s^2 2p^6 3s^2 3p^6 3d^{10}
4s^2 4p^6 4d^{10} 4f^{14} 5s^2 5p^6
5d^{10} 5f^0 5g^0 6s^2 6p^1

Radioactive Isotopes
Tl (at.wt.=204); ½life=3.56 years
with decay via β^- and K.
Radioactivity is induced.

Thorium   Th

Principal Quantum No.	7		Valence Electrons	$6d^2\ 7s^2$
Atomic No.	90		Acid-Base Property	Moderately basic
Atomic Wt.	232 △		Density	11.7 gm/ml
X-Ray Notation	Q		Crystal Structure	Cubic—face centered
Group	V B		Electro-negativity	1.3
Category	Actinon			
State	Solid		Electrical Conductance	0.055 micro-ohm
Origin	Natural			
No. of Protons	90		First Ioniza-tion Energy	/
No. of Electrons	90		Ionization Potential	/
Valence	+3, +4		Heat of Vaporization	/
Subshell Filling	f		Heat of Fusion	4.6 Kg-cal/gm-atom
Atomic Radius	/		Specific Heat	0.034 cal/gm/°C
Covalent Radius	1.65 Å		Thermal Conductance	/
Ionic Radius	1.14 Å (+3)		Boiling Point	3850° C
Atomic Vol.	19.9 W/D		Melting Point	1750° C
Core	Krypton			

Ground State Electron Config.
$1s^2\ 2s^2\ 2p^6\ 3s^2\ 3p^6\ 3d^{10}$
$4s^2\ 4p^6\ 4d^{10}\ 4f^{14}\ 5s^2\ 5p^6$
$5d^{10}\ 5f^0\ 5g^0\ 6s^2\ 6p^6\ 6d^2$
$7s^2$

Radioactive Isotopes
Th (at.wt.=232); ½life=1.4×10^{10} years with decay via α and γ; spontaneous fission.
Th (at.wt.=228); ½life=1.91 years with decay via β^-.
These isotopes occur naturally.

Thulium   Tm

Principal Quantum No.	6	Valence Electrons	$4f^{13}$
Atomic No.	69	Acid-Base Property	Slightly basic
Atomic Wt.	169.934	Density	9.05 gm/ml
X-Ray Notation	P	Crystal Structure	Hexagonal
Group	VI A	Electro-negativity	1.2
Category	Lanthanon		
State	Solid	Electrical Conductance	0.011 micro-ohm
Origin	Natural		
No. of Protons	69	First Ioniza-tion Energy	/
No. of Electrons	69	Ionization Potential	/
Valence	+3 (also, +2)	Heat of Vaporization	59 Kg-cal/gm-atom
Subshell Filling	f	Heat of Fusion	4.4 Kg-cal/gm-atom
Atomic Radius	1.74 Å	Specific Heat	0.038 cal/gm/°C
Covalent Radius	1.56	Thermal Conductance	/
Ionic Radius	0.95 Å(+3)	Boiling Point	1727° C
Atomic Vol.	18.1 W/D	Melting Point	1545° C
Core	Krypton		

Ground State Electron Config.
$1s^2\ 2s^2\ 2p^6\ 3s^2\ 3p^6\ 3d^{10}$
$4s^2\ 4p^6\ 4d^{10}\ 4f^{13}\ 5s^2\ 5p^6$
$5d^0\ 5f^0\ 5g^0\ 6s^2$

Radioactive Isotopes
Tm (at.wt.=170); ½life=127 days
with decay via β^-, γ and e^-.
Radioactivity is induced.

105

Tin   Sn (Stannum)

Principal Quantum No.	5		Valence Electrons	5p²
Atomic No.	50		Acid-Base Property	Amphoteric
Atomic Wt.	118.69			
X-Ray Notation	Q			
Group	IV A		Density	7.30 gm/ml
Category	Heavy transitional metal		Crystal Structure	Tetragonal
State	Solid		Electronegativity	1.8
Origin	Natural			
No. of Protons	50		Electrical Conductance	0.088 micro-ohm
No. of Electrons	50		First Ionization Energy	169 Kcal/gm-mole
Valence	+2, +4		Ionization Potential	7.3 ev
Subshell Filling	p		Heat of Vaporization	70 Kg-cal/gm-atom
Atomic Radius	1.62 Å		Heat of Fusion	1.72 Kg-cal/gm-atom
Covalent Radius	1.41 Å		Specific Heat	0.054 cal/gm/°C
Ionic Radius	1.12 Å(+2); 0.71 Å(+4)		Thermal Conductance	0.16 cal/cm²/cm/°C/sec
Atomic Vol.	16.3 W/D		Boiling Point	2270° C
Core	Krypton		Melting Point	231.9° C

Ground State Electron Config.
1s² 2s² 2p⁶ 3s² 3p⁶ 3d¹⁰
4s² 4p⁶ 4d¹⁰ 4f⁰ 5s² 5p²

Radioactive Isotopes
Sn (at.wt.=113); ½life=119 days
with decay via γ, e⁻, K and L.
Radioactivity is induced.

Titanium   Ti

Principal Quantum No.	4	Valence Electrons	3d²
Atomic No.	22	Acid-Base Property	Amphoteric
Atomic Wt.	47.90	Density	4.51 gm/ml
X-Ray Notation	N	Crystal Structure	Hexagonal
Group	IV B	Electronegativity	1.5
Category	Heavy transitional metal	Electrical Conductance	0.024 micro-ohm
State	Solid	First Ionization Energy	158 Kcal/gm-mole
Origin	Natural	Ionization Potential	6.8 ev
No. of Protons	22	Heat of Vaporization	106.5 Kg-cal/gm-atom
No. of Electrons	22	Heat of Fusion	3.7 Kg-cal/gm-atom
Valence	+2, +3, +4	Specific Heat	0.126 cal/gm/°C
Subshell Filling	d	Thermal Conductance	/
Atomic Radius	1.47 Å	Boiling Point	3260° C
Covalent Radius	1.36 Å	Melting Point	1668° C
Ionic Radius	0.90 Å(+2); 0.68 Å(+4)		
Atomic Vol.	10.6 W/D		
Core	Argon		

Ground State Electron Config.   Radioactive Isotopes
$1s^2\ 2s^2\ 2p^6\ 3s^2\ 3p^6\ 3d^2\ 4s^2$ None.

Tungsten   W (Wolfram)

Principal Quantum No.	6	Valence Electrons	$5d^4$
Atomic No.	74	Acid-Base Property	Moderately acidic
Atomic Wt.	183.85	Density	19.3 gm/ml
X-Ray Notation	P	Crystal Structure	Cubic—body centered
Group	VI B	Electronegativity	1.7
Category	Heavy transitional metal	Electrical Conductance	0.181 micro-ohm
State	Solid	First Ionization Energy	184 Kcal/gm-mole
Origin	Natural	Ionization Potential	8.0 ev
No. of Protons	74	Heat of Vaporization	185 Kg-cal/gm-atom
No. of Electrons	74	Heat of Fusion	8.05 Kg-cal/gm-atom
Valence	+6 (also, +2, +3, +4, +5)	Specific Heat	0.032 cal/gm/°C
Subshell Filling	d	Thermal Conductance	0.40 cal/cm²/cm/°C/sec
Atomic Radius	1.39 Å	Boiling Point	5930° C
Covalent Radius	/	Melting Point	3410° C
Ionic Radius	0.64 Å(+4); 0.68 Å(+6)		
Atomic Vol.	9.53 W/D		
Core	Krypton		

Ground State Electron Config.
$1s^2\ 2s^2\ 2p^6\ 3s^2\ 3p^6\ 3d^{10}$
$4s^2\ 4p^6\ 4d^{10}\ 4f^{14}\ 5s^2\ 5p^6$
$5d^4\ 5f^0\ 5g^0\ 6s^2$

Radioactive Isotopes
W (at.wt.=185); ½life=73 days
with decay via β^- and γ.
Radioactivity is induced.

Uranium   U

Principal Quantum No.	7	Valence Electrons	5f³ 6d¹
Atomic No.	92	Acid-Base Property	Amphoteric
Atomic Wt.	238.041 △	Density	19.07 gm/ml
X-Ray Notation	Q	Crystal Structure	Orthorhombic
Group	VII B		
Category	Actinon	Electro-negativity	1.7
State	Solid	Electrical Conductance	0.034 micro-ohm
Origin	Natural		
No. of Protons	92	First Ioniza-tion Energy	/
No. of Electrons	92	Ionization Potential	4 ev
Valence	+3, +4, +6 (also, +2, +5)	Heat of Vaporization	110 Kg-cal/gm-atom
Subshell Filling	f	Heat of Fusion	2.7 Kg-cal/gm-atom
Atomic Radius	/	Specific Heat	0.028 cal/gm/°C
Covalent Radius	1.42 Å	Thermal Conductance	0.064 cal/cm²/cm/°C/sec
Ionic Radius	1.11 Å(+3); 0.89 Å(+4)	Boiling Point	3818° C
		Melting Point	1132° C
Atomic Vol.	12.5 W/D		
Core	Krypton		

Ground State Electron Config.
1s² 2s² 2p⁶ 3s² 3p⁶ 3d¹⁰
4s² 4p⁶ 4d¹⁰ 4f¹⁴ 5s² 5p⁶
5d¹⁰ 5f³ 5g⁰ 6s² 6p⁶ 6d¹
7s²

Radioactive Isotopes

U (at.wt.=238); ½life=4.5×10⁹ years with decay via α and γ; spontaneous fission.

U (at.wt.=234); ½life=2.5×10⁵ years with decay via α and γ; spontaneous fission.

U (at.wt.=235); ½life=7.1×10⁸ years with decay via α and γ; spontaneous fission.

The above three isotopes occur naturally.

U (at.wt.=233); ½life=1.6×10⁵ years with decay via α and γ.

Radioactivity induced in U-233 only.

Vanadium
V

Principal Quantum No.	4	Valence Electrons	3d³	
Atomic No.	23	Acid-Base Property	Amphoteric	
Atomic Wt.	50.942	Density	6.1 gm/ml	
X-Ray Notation	N	Crystal Structure	Cubic—body centered	
Group	V B			
Category	Heavy transitional metal	Electronegativity	1.5	
State	Solid	Electrical Conductance	0.024 micro-ohm	
Origin	Natural	First Ionization Energy	158 Kcal/gm-mole	
No. of Protons	23	Ionization Potential	6.7 ev	
No. of Electrons	23	Heat of Vaporization	106.5 Kg-cal/gm-atom	
Valence	+2, +3, +4, +5	Heat of Fusion	3.7 Kg-cal/gm-atom	
Subshell Filling	d	Specific Heat	0.126 cal/gm/°C	
Atomic Radius	1.47 Å	Thermal Conductance	/	
Covalent Radius	1.36 Å	Boiling Point	3450° C	
Ionic Radius	0.90 Å(+2); 0.68 Å(+4)	Melting Point	1900° C	
Atomic Vol.	10.6 W/D			
Core	Argon			

Ground State Electron Config.
1s² 2s² 2p⁶ 3s² 3p⁶ 3d³ 4s²

Radioactive Isotopes
None.

Xenon   Xe

Principal Quantum No.	5	Valence Electrons	5p^6
Atomic No.	54	Acid-Base Property	/
Atomic Wt.	131.30	Density	3.06 gm/ml
X-Ray Notation	Q	Crystal Structure	Cubic—face centered
Group	Inert gas	Electro-negativity	/
Category	Inert gas		
State	Gas	Electrical Conductance	/
Origin	Natural		
No. of Protons	54	First Ionization Energy	280 Kcal/gm-mole
No. of Electrons	54	Ionization Potential	12.1 ev
Valence	0	Heat of Vaporization	302 Kg-cal/gm-atom
Subshell Filling	p	Heat of Fusion	0.55 Kg-cal/gm-atom
Atomic Radius	/	Specific Heat	/
Covalent Radius	2.09 Å	Thermal Conductance	0.0001 cal/cm^2/cm/°C/sec
Ionic Radius	/	Boiling Point	108° C
Atomic Vol.	42.9 W/D	Melting Point	111.9° C
Core	Krypton		

Ground State Electron Config.
1s^2 2s^2 2p^6 3s^2 3p^6 3d^{10}
4s^2 4p^6 4d^{10} 4f^0 5s^2 5p^6

Radioactive Isotopes
None.

Ytterbium   Yb

Principal Quantum No.	6		Valence Electrons	$4f^{14}\ 6s^2$
Atomic No.	70		Acid-Base Property	Slightly basic
Atomic Wt.	173.04		Density	6.98 gm/ml
X-Ray Notation	P		Crystal Structure	Cubic—face centered
Group	VII A			
Category	Lanthanon		Electro-negativity	1.1
State	Solid		Electrical Conductance	0.035
Origin	Natural			
No. of Protons	70		First Ioniza-tion Energy	143 Kcal/gm-mole
No. of Electrons	70		Ionization Potential	6.2 ev
Valence	+3 (also, +2)		Heat of Vaporization	38 Kg-cal/gm-atom
Subshell Filling	f		Heat of Fusion	1.8 Kg-cal/gm-atom
Atomic Radius	1.92 Å		Specific Heat	0.035 cal/gm/°C
Covalent Radius	1.70 Å		Thermal Conductance	/
Ionic Radius	1.13 Å(+2); 0.94 Å(+3)		Boiling Point	1427° C
Atomic Vol.	24.8 W/D		Melting Point	824° C
Core	Krypton			

Ground State Electron Config.
$1s^2\ 2s^2\ 2p^6\ 3s^2\ 3p^6\ 3d^{10}$
$4s^2\ 4p^6\ 4d^{10}\ 4f^{14}\ 5s^2\ 5p^6$
$5d^0\ 5f^0\ 5g^0\ 6s^2$

Radioactive Isotopes
Yb (at.wt.=175); ½life=4.2 days
with decay via β^- and γ.
Yb (at.wt.=169); ½life=31 days
with decay via γ, e^- and K.
Radioactivity is induced.

112

Yttrium   Y

Principal Quantum No.	5	Valence Electrons	$4d^1\ 5s^2$
Atomic No.	39	Acid-Base Property	Moderately basic
Atomic Wt.	88.905	Density	4.47 gm/ml
X-Ray Notation	Q	Crystal Structure	Hexagonal
Group	III B		
Category	Heavy transitional metal	Electronegativity	1.3
State	Solid	Electrical Conductance	0.019 micro-ohm
Origin	Natural	First Ionization Energy	152 Kcal/gm-mole
No. of Protons	39		
No. of Electrons	39	Ionization Potential	6.6 ev
Valence	+3	Heat of Vaporization	93 Kgcal/gm-atom
Subshell Filling	d	Heat of Fusion	2.7 Kg-cal/gm-atom
Atomic Radius	1.80 Å	Specific Heat	0.071 cal/gm/°C
Covalent Radius	1.62 Å	Thermal Conductance	0.035 cal/cm²/cm/°C/sec
Ionic Radius	0.93 Å(+3)	Boiling Point	2927° C
Atomic Vol.	19.8 W/D	Melting Point	1509° C
Core	Krypton		

Ground State Electron Config.
$1s^2\ 2s^2\ 2p^6\ 3s^2\ 3p^6\ 3d^{10}$
$4s^2\ 4p^6\ 4d^1\ 4f^0\ 5s^2$

Radioactive Isotopes
Y (at.wt.=90); ½life=64 hours
with decay via β^- and e^-.
Radioactivity is induced.

Zinc   Zn

Principal Quantum No.	4		Valence Electrons	$4s^2$
Atomic No.	30		Acid-Base Property	Amphoteric
Atomic Wt.	65.37		Density	7.14 gm/ml
X-Ray Notation	N		Crystal Structure	Hexagonal
Group	II B		Electro-negativity	1.6
Category	Metal			
State	Solid		Electrical Conductance	0.167 micro-ohm
Origin	Natural			
No. of Protons	30		First Ioniza-tion Energy	216 Kcal/gm-mole
No. of Electrons	30		Ionization Potential	9.4 ev
Valence	+2		Heat of Vaporization	27.4 Kg-cal/gm-atom
Subshell Filling	s		Heat of Fusion	1.76 Kg-cal/gm-atom
Atomic Radius	1.38 Å		Specific Heat	0.0915 cal/gm/°C
Covalent Radius	1.31 Å		Thermal Conductance	0.27 cal/cm²/cm/°C/sec
Ionic Radius	0.74 Å(+2)		Boiling Point	906° C
Atomic Vol.	9.2 W/D		Melting Point	419.5° C
Core	Argon			

Ground State Electron Config.
$1s^2\ 2s^2\ 2p^6\ 3s^2\ 3p^6\ 3d^{10}$
$4s^2$

Radioactive Isotopes
Zn (at.wt.=65); ½life=245 days
with decay via β^+, γ and K.
Radioactivity is induced.

Zirconium   Zr

Principal Quantum No.	5		Valence Electrons	$4d^2$ $5s^2$
Atomic No.	40		Acid-Base Property	Amphoteric
Atomic Wt.	91.22		Density	6.49 gm/ml
X-Ray Notation	Q		Crystal Structure	Hexagonal
Group	IV B			
Category	Heavy transitional metal		Electronegativity	1.4
State	Solid		Electrical Conductance	0.024 micro-ohm
Origin	Natural		First Ionization Energy	160 Kcal/gm-mole
No. of Protons	40		Ionization Potential	7.0 ev
No. of Electrons	40		Heat of Vaporization	120 Kg-cal/gm-atom
Valence	+4 (also, +2, +2)		Heat of Fusion	4.0 Kg-cal/gm-atom
Subshell Filling	d		Specific Heat	0.066 cal/gm/°C
Atomic Radius	1.60 Å		Thermal Conductance	/
Covalent Radius	1.48 Å		Boiling Point	3580° C
Ionic Radius	0.80 Å(+4)		Melting Point	1852° C
Atomic Vol.	14.1 W/D			
Core	Krypton			

Ground State Electron Config.
$1s^2$ $2s^2$ $2p^6$ $3s^2$ $3p^6$ $3d^{10}$
$4s^2$ $4p^6$ $4d^2$ $4f^0$ $5s^2$

Radioactive Isotopes
Zr (at.wt.=95); ½life=65 days
with decay via β^-, γ and e⁻.
Zr (at.wt.=93); ½life=9×10⁵ years
with decay via β^- and γ.
Radioactivity is induced.

115

The following diagrams demonstrate the types of crystal structures observed among the elements:

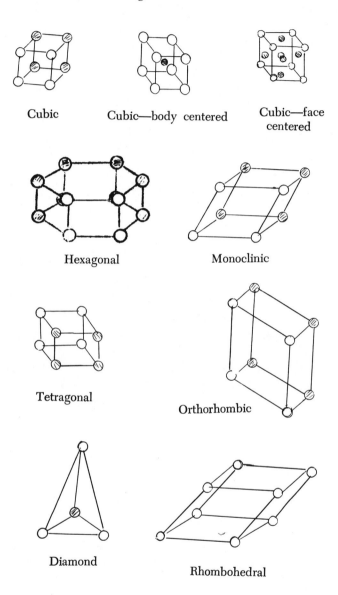

Cubic   Cubic—body centered   Cubic—face centered

Hexagonal   Monoclinic

Tetragonal

Orthorhombic

Diamond

Rhombohedral

TABLE 1   Percentage composition of elements in the earth's crust.

Oxygen	46.7%
Silicon	27.7%
Aluminum	8.1%
Iron	5.1%
Calcium	3.5%
Sodium	2.7%
Potassium	2.5%
Magnesium	2.1%
Chlorine	0.06%
Fluorine	0.03%
Bromine	0.001%
All others	2.109%

TABLE 2   Percentage composition of elements in dry air.

Nitrogen	78.03%
Oxygen	20.99%
Argon	0.94%
Hydrogen	0.011%
Neon	0.0015%
Helium	0.0005%
Krypton	0.0001%
Xenon	0.00009%
Radon	Trace

Also, Carbon dioxide, 0.031%

TABLE 3   General tendencies of some periodic properties.

Schema of the table of periodic classification of the elements.

inert gases

TABLE 3   General tendencies of some periodic properties. (*continued*)

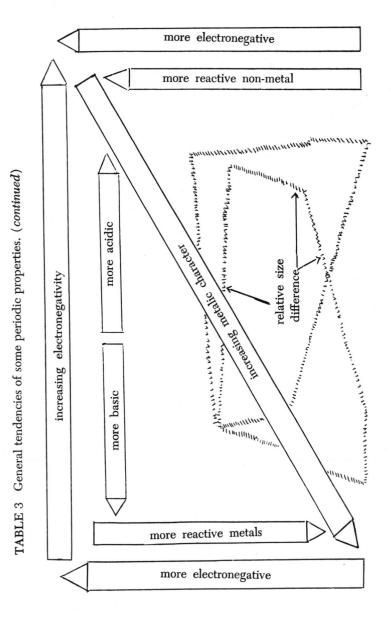

more electronegative

more reactive non-metal

more acidic

more basic

increasing electronegativity

increasing metallic character

relative size difference

more reactive metals

more electronegative

119

TABLE 4 Physical constants.

Avogadro's number...$N = 6.0254 \times 10^{26}$Kg/mole
Bohr magneton...$\mu B = 9.273 \times 10^{-24}$Am
Electron charge...$e = 1.6019 \times 10^{-19}$C
Electron charge...$e = 4.8022 \times 10^{-10}$esu
Electron mass energy equivalent...$mc^2 = 0.51097$ MeV
Atomic mass unit (energy)...$E = 931.34$ MeV
Electron mass...$m = 9.1072 \times 10^{-31}$Kg
Mass of atom (unit weight)...$M = 1.6600 \times 10^{-27}$Kg
Mass of proton...$M_p = 1.6722 \times 10^{-27}$Kg
Planck's action constant...$h = 6.6238 \times 10^{-34}$Js
Proton mass:electron mass (ratio)...$M_p/m = 1836.1$
Specific electronic charge...$e/m = 1.7598 \times 10^{11}$C/Kg
Specific proton charge...$e/M_p = 9.5795 \times 10^7$C/Kg
Velocity of light...$c = 299,790$ Km/sec
Velocity of light^2...$c^2 = 8.9874 \times 10^{10}$Km^2/s^2

TABLE 5 Conversion factors

1 Angstrom (Å) $= 10^{-10}$m
1 millimicron (mμ) $= 10^{-9}$m
1 micron (μ) $= 10^{-6}$m
1 millimeter (mm) $= 10^{-3}$m
1 centimeter (cm) $= 10^{-2}$m
1 meter (m) $= 100$ cm
1 kilometer (km) $= 1000$ m

1 inch (in) $= 2.54$ cm
1 foot (ft) $= 30.48$ cm
1 mile (mi) $= 1.609$ km
1 centimeter $= 0.3937$ in
1 meter $= 39.37$ in
1 kilometer $= 0.6214$ mi

Temperature conversion
°K (or °A) $= $°C$ + 273$
°F $= ($°C$\times 9/5) - 32$
°C $= ($°F$ - 32) \times 5/9$

TABLE 6 Maximum electron populations in main energy levels.

	1	2	3	4	5	6	7
Principal quantum no., n.	1	2	3	4	5	6	7
X-ray notation	K	L	M	N	O	P	Q
Maximum no. of electrons in levels, designated by sublevel=$2n^2$.	s2	s2 p6	s2 p6 d10	s2 p6 d10 f14	s2 p6 d10 f14 g18	s2 p6 d10 f14 g18 h22	s2 p6 d10 f14 g18 h48
Totals	2	8	18	32	50	72	98
Atomic numbers of elements in these periods.	1+2	3–10	11–18	19–36	37–54	55–86	87–103

TABLE 7   Filling order of main energy levels.

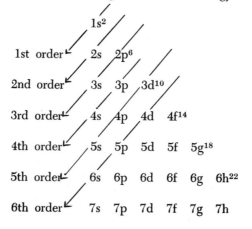

1st order   1s² 2s 2p⁶

2nd order   3s 3p 3d¹⁰

3rd order   4s 4p 4d 4f¹⁴

4th order   5s 5p 5d 5f 5g¹⁸

5th order   6s 6p 6d 6f 6g 6h²²

6th order   7s 7p 7d 7f 7g 7h

TABLE 8   Activity series.

Reductant	Oxidant	Potential
Li	Li$^+$	+3.95
Rb	Rb$^+$	+2.93
K	K$^+$	+2.93
Cs	Cs$^+$	+2.92
Ba	Ba^{++}	+2.90
Sr	Sr^{++}	+2.89
Ca	Ca^{++}	+2.87
Na	Na$^+$	+2.71
Mg	Mg^{++}	+2.37
Be	Be^{++}	+1.85
Al	Al^{+++}	+1.66
Mn	Mn^{++}	+1.18
Zn	Zn^{++}	+0.76
Cr	Cr^{+++}	+0.74
Ga	Ga^{+++}	+0.53
S$^=$	S	+0.48
Fe	Fe^{++}	+0.44
Cd	Cd^{++}	+0.40
Tl	Tl$^+$	+0.38
Co	Co^{++}	+0.28
Ni	Ni^{++}	+0.25
Sn	Sn^{++}	+0.14
Pb	Pb^{++}	+0.13
H	H$^+$	0.00
Sn^{++}	Sn^{++++}	−0.15
Cu	Cu^{++}	−0.34
I$^-$	I$_2$	−0.54
Fe^{++}	Fe^{+++}	−0.77
Hg	Hg$_2$$^{++}$	−0.79
Ag	Ag$^+$	−0.80
Hg$_2$$^{++}$	Hg^{++}	−0.92
Br$^-$	Br$_2$	−1.07
Cl$^-$	Cl$_2$	−1.36
Au	Au$^+$	−1.68
F$^-$	F$_2$	−2.65

increasing tendency toward oxidized state

increasing tendency toward reduced state

123

Symbols

The symbols that define the method of decay and the particle being emitted in this process, observed in the radioisotopes, are explained as follows:

α = an alpha particle
β^+ = a positron
β^- = a beta particle
γ = gamma ray
L = capture of electron populating the L shell
K = capture of electron populating the K shell
e^- = internal conversion of an electron
SF = spontaneous fission

The Greek Alphabet

A	α	Alpha (a)
B	β	Beta (b)
Γ	γ	Gamma (g)
Δ	δ	Delta (d)
E	ϵ	Epsilon (e)
Z	ζ	Zeta (z)
H	η	Eta (h)
Θ	θ	Theta (th)
I	ι	Iota (i)
K	κ	Kappa (k)
Λ	λ	Lambda (l)
M	μ	Mu (m)
N	ν	Nu (n)
Ξ	ξ	Xi (x)
O	o	Omicron (o)
Π	π	Pi (p)
P	ρ	Rho (r)
Σ	σ	Sigma (s)
T	τ	Tau (t)
Υ	υ	Upsilon (u)
Φ	ϕ	Phi (hp)
X	χ	Chi (ch)
Ψ	ψ	Psi (ps)
Ω	ω	Omega (o)

TABLE OF PERIODIC CLASSIFICATION OF THE ELEMENTS

Principal quantum number, n (period)									GROUP
7	6	7	6	5	4	3	2	1	
Actinons ′	Lanthanons ″	87 Fr	55 Cs	37 Rb	19 K	11 Na	3 Li	1 H	I A
		88 Ra	56 Ba	38 Sr	20 Ca	12 Mg	4 Be		II A
		89 (Act)′	57 (L)″	39 Y	21 Sc				III B
89 Ac	57 La	72 Hf		40 Zr	22 Ti				IV B
90 Th	58 Ce	73 Ta		41 Nb	23 V				V B
91 Pa	59 Pr	74 W		42 Mo	24 Cr				VI B
92 U	60 Nd	75 Re		43 Tc	25 Mn				VII B
93 Np	61 Pm	76 Os		44 Ru	26 Fe				VIII B
94 Pu	62 Sm	77 Ir		45 Rh	27 Co				
95 Am	63 Eu	78 Pt		46 Pd	28 Ni				
96 Cm	64 Gd	79 Au		47 Ag	29 Cu				I B
97 Bk	65 Tb	80 Hg		48 Cd	30 Zn				II B
98 Cf	66 Dy			49 In	31 Ga	13 Al	5 B		III A
99 Es	67 Ho			50 Sn	32 Ge	14 Si	6 C		IV A
100 Fm	68 Er			51 Sb	33 As	15 P	7 N		V A
101 Md	69 Tm			52 Te	34 Se	16 S	8 O		VI A
102 No	70 Yb			53 I	35 Br	17 Cl	9 F		VII A
103 Lw	71 Lu			54 Xe	36 Kr	18 Ar	10 Ne	2 He	VIII A
Q	P	Q	P	O	N	M	L	K	
X-ray Notation									